Superscalar Microprocessor Design

Prentice Hall Series in Innovative Technology

Dennis R. Allison, David J. Farber, and Bruce D. Shriver   *Series Advisors*

Superscalar Microprocessor Design

Mike Johnson

Advanced Micro Devices

Prentice Hall, Englewood Cliffs, New Jersey 07632

Library of Congress Cataloging-in-Publication Data

Johnson, William.
  Superscalar microprocessor design / William Johnson.
      p.   cm. -- (Prentice Hall series in innovative technology)
  Includes bibliographical references (p.).
  Includes index.
  ISBN 0-13-875634-1
  1. Microprocessors--Design and construction. 2. Reduced
instruction set computers.   I. Title. II. Series.
TK7895.M5J64 1991
621.39'16--dc20   90-14270
                                                            CIP

Editorial/production supervision: Harriet Tellem
Cover design: Karen Stephens
Cover art: © 1990 M. C. Escher Heirs/Cordon Art, Baarn, Holland
Manufacturing buyers: Kelly Behr/Susan Brunke

© 1991 by Prentice-Hall, Inc.
A division of Simon & Schuster
Englewood Cliffs, New Jersey 07632

Am29000 is a trademark of Advanced Micro Devices, Inc.

386, i486, and i860 are trademarks of Intel, Inc.

RISC System/6000 is a trademark of International Business Machines, Inc.

R2000, R2010, and pixie are trademarks of MIPS, Inc.

The publisher offers discounts on this book when ordered
in bulk quantities. For more information, write:

Special Sales/College Marketing
Prentice-Hall, Inc.
College Technical and Reference Division
Englewood Cliffs, New Jersey 07632

Printed in the United States of America
10 9 8 7 6 5 4 3 2 1

ISBN 0-13-875634-1

Prentice-Hall International (UK) Limited, *London*
Prentice-Hall of Australia Pty. Limited, *Sydney*
Prentice-Hall Canada Inc., *Toronto*
Prentice-Hall Hispanoamericana, S.A., *Mexico*
Prentice-Hall of India Private Limited, *New Delhi*
Prentice-Hall of Japan, Inc., *Tokyo*
Simon & Schuster Asia Pte. Ltd., *Singapore*
Editora Prentice-Hall do Brasil, Ltda., *Rio de Janeiro*

To my wife
Mary
and my sons
Kevin and Brian

Contents

Contents

Figures

Figures

Tables

Preface

The term *superscalar* describes a computer implementation that improves performance by concurrent execution of scalar instructions–the type of instructions typically found in general-purpose microprocessors. Because the majority of existing microprocessor applications are targeted toward scalar computation, superscalar microprocessors are the next logical step in the evolution of microprocessors. Using today's semiconductor processing technology, a single processor chip can incorporate high-performance techniques that were once applicable only to large-scale, scientific processors. However, many of the techniques applied to large-scale processors are either inappropriate for scalar computation or too expensive to be applied to microprocessors. This book brings together, for the first time, a comprehensive survey of the issues surrounding superscalar microprocessors.

Up to this point, the computer professional or student interested in superscalar techniques has been obliged to glean information from a wide variety of published literature. This approach presents several difficulties. First, much of the existing literature is not very accessible, because it is directed toward computer researchers. There is typically little background information to aid the unfamiliar reader. Second, most of the relevant research concerns the performance of scientific, numerically intensive applications, rather than the general-purpose, user-oriented applications that are in more widespread use. The differences between scientific and general-purpose applications are often not described explicitly in the literature, even though these differences affect processor design in important ways. Third, there is a sharp division in the literature between hardware and software techniques, with little attention given to the synthesis of hardware and software. This gives the appearance of a dualism–hardware-intensive versus software-intensive approaches–where none exists. Finally, the published literature often deals with idealizations and ignores real-world concerns such as cost-effectiveness and software compatibility. One hardly knows how to choose the best approaches from among all the alternatives.

This book is intended as a technical tutorial and introduction for engineers and computer scientists as well as a graduate-level text for students who have a strong background in computer architecture. It is oriented toward the general-purpose application of superscalar microprocessors rather than scientific applications or scientific processors, because the reader is more likely to encounter general-purpose applications in a professional career, either as a designer or user. Furthermore, this book concentrates on reduced-instruction-set (RISC) processors and assumes that readers have some familiarity with RISC design techniques. Focusing on RISC architectures helps to simplify the technical discussion by avoiding issues related to complex instruction sets. Despite the focus on the general-purpose application of RISC processors, however, this book does point out considerations related to scientific applications and complex-instruction-set (CISC) processors, where appropriate. The Appendix describes issues related to designing a superscalar version of a popular CISC architecture–the Intel 386™–and discusses methods of approaching such a design.

This book is organized as follows:

Chapter 1 describes a progression of techniques for increasing microprocessor performance in general-purpose applications. Superscalar processors follow quite naturally in this progression and are in a sense an extension of pipelined scalar processors. This chapter sets the tone and scope of the book.

Chapter 2 continues to develop superscalar architectures as an extension of pipelined scalar architectures. It introduces basic superscalar principles and describes basic limits to performance. This chapter also describes techniques for improving performance, such as out-of-order issue and register renaming, and relates superscalar techniques to other ideas for high-performance processors.

Chapters 3-8 form an extensive superscalar hardware evaluation that explores a number of implementation alternatives. These chapters develop a superscalar model of the MIPS, Inc., R2000™ processor and use this model to explore the cost, complexity, and performance benefits of various superscalar hardware components. In contrast to most of the previous work on superscalar and related processors, this analysis is based on the simulation of large, real, general-purpose software applications. These chapters are organized as follows:

Chapter 3 develops a superscalar simulation model and introduces concepts that are used in Chapters 4-8 to evaluate this model.

Chapter 4 uses the superscalar model to show that it is difficult to fetch and decode instructions at an adequate rate to satisfy the superscalar execution unit. This chapter also describes techniques for overcoming the instruction-fetch limitation. Even though these limitations can be reduced, instruction fetching presents the most severe limit to superscalar performance. This observation affects all other design decisions. For example, complicated techniques to improve the parallelism of hardware are unwarranted, because instruction-fetch limitations prevent most programs from exploiting this parallelism.

Chapter 5 describes exception recovery, which is important both in its own right and as an important element in efficient instruction fetching and decoding.

Chapter 6 describes the implementation of register-based dataflow: that is, the dependency analysis and data interconnections required to route register-based operands to their respective operations. This chapter explores a large number of dependency algorithms and shows that many alternatives are both more complex and provide less performance than a straightforward approach.

Chapter 7 examines mechanisms for initiating the execution of more that one instruction per cycle, using out-of-order issue. It examines the advantages and disadvantages of distributed instruction issue as compared to centralized instruction issue.

Chapter 8 describes the implementation of memory-based dataflow. This chapter has a similar objective to that of Chapter 6, except that it concerns memory-based operands accessed via loads and stores (using the RISC model) rather than register-based operands.

Chapter 9 summarizes the results of Chapters 4-8, examining the superscalar processor in light of design complexity. The complexity of a superscalar design motivates an investigation of software techniques which might help simplify the hardware.

Chapter 10 describes basic techniques and algorithms used by software to optimize code for a superscalar processor. These basic optimizations deal primarily with code that does not contain branches. This chapter is not a rigorous survey of optimization techniques, but rather is aimed at understanding the capabilities and limitations of these techniques. It demonstrates that software optimizations are important even with the most aggressive hardware designs.

Chapter 11 explores software optimizations that can be performed in code that contains branches. These types of optimizations are necessary if software techniques, used with simple hardware, are to achieve the performance of complex hardware techniques. However, the nature of these optimizations do not settle the argument for or against complex hardware. This chapter exposes the real dualism between hardware-intensive and software-intensive approaches: speculative, out-of-order issue versus branch-related optimizations.

Chapter 12 reexamines the topics raised in previous chapters in the context of microprocessor evolution. It shows how different types of superscalar designs result from different assumptions about costs and user requirements. This chapter provides readers with important criteria for judging and evaluating different designs and proposals.

This book is rare among technical publications, because it charts a coming technical revolution instead of documenting the past. Because it focuses on real designs, real applications, and real-world trade-offs, this material is recommended to anyone wanting to stay abreast of the rapid progress of microprocessor technology–especially anyone who is contemplating designing, funding, or evaluating a superscalar microprocessor.

Much of the research for this book was performed at Stanford University. I am indebted to Mark Horowitz and Mike Smith for many important ideas and for steering me in the right direction when I got off track. I also had many long conversations with Monica Lam which were invaluable in helping me understand the trade-offs between hardware and software techniques. David Witt, a colleague at AMD, and Mike Smith extensively reviewed an early manuscript and provided many helpful comments. Monica Lam, Scott McMahon, Paul Lum, and Dennis Allison also reviewed all or various parts of this book and helped me to improve its quality. I owe a special debt of gratitude to Advanced Micro Devices, Inc., which has supported my research, provided the resources to complete this book, and–most important–given me the rare opportunity to gain insight into the workings of microprocessor technology, from product conception, through development, all the way to seeing products that serve users. This experience has given me much needed confidence to put forth the ideas in this book, as speculative as they sometimes are. Finally, I wish to thank my wife, Mary and sons, Kevin and Brian. They have patiently put up with many busy evenings and weekends while I completed this book. To them, I promise: no more large writing projects, at least for a while!

Chapter 1

Beyond Pipelining, CISC, and RISC

This book is about improving the performance of microprocessors. Microprocessors, by definition, must be implemented on one, or a very small number of, semiconductor chips. Semiconductor technology provides ever-increasing circuit densities and speeds for implementing a microprocessor, but the interconnection with the microprocessor's memory is quite constrained by packaging technology. Though on-chip interconnections are extremely cheap, off-chip connections are very expensive; often the processor's package and pins are more expensive than the processor chip itself. Any technique intended to improve microprocessor performance must take advantage of increasing circuit densities and speeds while remaining within the constraints of packaging technology and the physical separation between the processor and its memory. At the same time, though increasing circuit densities provide a path to ever-more-complex designs, the operation of the microprocessor must remain simple and clear enough that users can understand how to use it. In some ways, this is the most important constraint.

This book is also about improving the performance of general-purpose applications. I have never liked the term "general purpose" because it is so vague that it is almost meaningless. But in this book, my intent precisely is to be vague. The outstanding characteristic of general-purpose applications is that they are hard to characterize. It is hard to identify properties of the applications that reveal how to improve performance. Some applications–ones that we will not deal with very much here–lend themselves quite easily to specialized techniques. For example, some signal-processing applications are performed well by systolic arrays, and some scientific computations are performed well by vector processors. We are more interested in applications in everyday use and will use a criterion that frustrates many design ideas: a technique has questionable value if it does not increase the performance of a reasonably large sample of applications.

We are interested in improving the performance of processors that behave, conceptually, as shown in Figure 1-1. An application program comprises a group of instructions. The processor fetches and executes instructions in some sequence ($i0$, $i1$, and so on in Figure 1-1). There are several steps involved in the execution of a single instruction: fetching the instruction, decoding it, assembling its operands, performing the operations specified by the instruction, and writing the results of the instruction to storage. The execution of instructions is controlled by a periodic clock signal; the period of the clock signal is the processor cycle time. These concepts should be trivial to anyone hoping to understand rest of the material in this book. Our goal in this introductory chapter is to use this simple concept to develop an historical progression of processors, each having improved performance over prior processors, and to show how superscalar processors fit naturally into this progression.

The time taken by a processor to complete a program is determined by three factors:

- the number of instructions required to execute the program
- the average number of processor cycles required to execute an instruction
- the processor cycle time

Processor performance is improved by reducing the time taken, which dictates reducing one or more of these factors. This is easily said, but there are many technical constraints to reducing these factors, and the constraints change as technology develops. The constraints that apply during the design of a given architecture do not necessarily apply during the design of the next, meaning that the later architecture

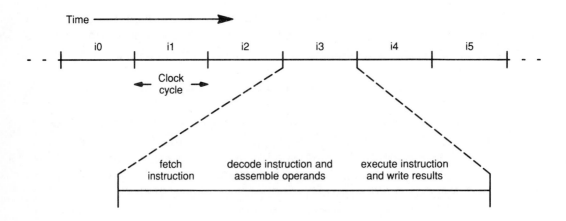

Figure 1-1. The Steps of Instruction Execution

has an opportunity to improve performance in ways that were not previously anticipated. This is why architectures evolve and new architectures replace older ones.
We can illustrate this by showing the evolution of microprocessors in idealized
terms. Ultimately, we will see how superscalar processors fit naturally into this progression.

Referring to Figure 1-1, an obvious way to increase performance is by overlapping the steps of different instructions, using a technique called *pipelining* [Kogge
1981]. To pipeline instructions, the various steps of instruction execution are performed by independent units called *pipeline stages*. Pipeline stages are separated by
clocked registers (or latches). The steps of different instructions are executed independently in different pipeline stages, as shown in Figure 1-2. The result of each
pipeline stage is communicated to the next pipeline stage via the register between the
stages. Pipelining reduces the average number of cycles required to execute an instruction, though not the total amount of time required to execute an instruction, by
permitting the processor to handle more than one instruction at a time. This is done
without increasing the processor cycle time appreciably. As Figure 1-2 illustrates,
pipelining reduces the average number of cycles per instruction by as much as a factor of three, in comparison to the timing in Figure 1-1.

The steps shown in Figure 1-2 do not necessarily take the same amount of time.
In particular, during the development of early microprocessors, instructions took a
long time to fetch, compared to the execution time, as shown in Figure 1-3. This
motivated the development of complex-instruction, or CISC, processors. (The acronym CISC stands for *complex-instruction-set computer*, so the term *CISC processor*

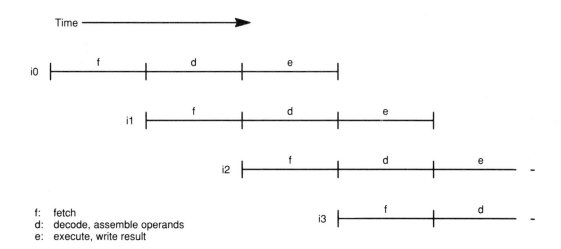

Figure 1-2. Pipelining Processor Instructions

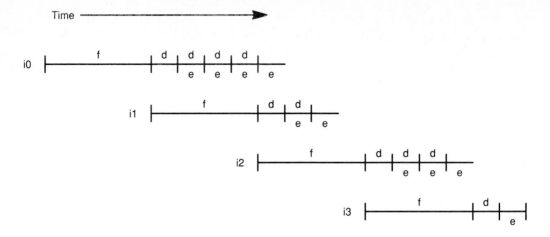

Figure 1-3. Pipelining in a Fetch-Limited (CISC) Processor

expands into the redundant phrase *complex-instruction-set-computer processor*. I find it easier to live with this redundancy than to use the acronym CISC by itself.) CISC processors were based on the observation that, given the available technology, the number of cycles per instruction was determined mostly by the number of cycles taken to fetch the instruction. To improve performance, the two principal goals of the CISC architecture were to reduce the number of instructions and to encode these instructions densely. It was acceptable to accomplish these goals by increasing the average number of cycles taken to decode and execute an instruction, because, using pipelining, the decode and execution cycles could be mostly overlapped with instruction fetch, as shown in Figure 1-3. With this set of assumptions, CISC processors evolved densely encoded instructions at the expense of decode and execution time inside the processor. Multiple-cycle instructions reduced the overall number of instructions, and thus reduced the overall execution time because they reduced the instruction-fetch time.

But, in the late 1970s and early 1980s, memory and packaging technology changed rapidly. High-pin-count packages made possible the design of advanced memory interfaces that no longer had quite the same fetch limitations as applied when CISC processors evolved. Memory densities and speeds increased to the point where high-speed local memories–called *caches* [Smith 1982]–could be implemented near the processor. Figure 1-4 shows the instruction timing of the CISC processor of Figure 1-3, but where instructions are fetched more quickly. The execution time is shorter than in Figure 1-3, but now performance is limited by the decode and execution time that was previously hidden within the instruction fetch time.

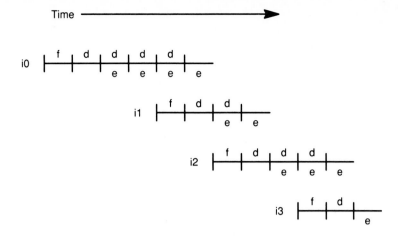

Figure 1-4. A CISC Processor Without Memory Limitations

Now, the number of instructions does not affect performance as much as the average number of cycles taken to execute an instruction.

The improvement in memory and packaging technology, to the point where instruction fetching did not take much longer than instruction execution, motivated the development of reduced-instruction, or RISC, processors (the acronym RISC stands for *reduced-instruction-set computer*). Figure 1-5 shows the timing of instruction execution in a RISC processor. To improve performance, the principle goal of a RISC architecture is to reduce the number of cycles taken to execute an instruction, allowing some increase in the total number of instructions [Hennessy 1986]. The trade-off between cycles per instruction and the number of instructions is not one to one: compared to CISC processors, RISC processors reduce the number of cycles per instruction by factors of roughly three to five, while they increase the number of instructions by 30% to 50%. Also, because the RISC instruction set allows access to primitive hardware operations, an optimizing compiler is effectively able to reduce the number of instructions performed [Auslander and Hopkins 1982, Chow 1983].

RISC processors have been characterized by some as a return to the basic, rudimentary architectures that were developed very early in the evolution of computers. However, early processors were simple because technology was relatively primitive. RISC processors are simple because simplicity yields better performance. Relative to CISC processors, RISC processors depend heavily on advanced memory technology, advanced packaging technology, and advanced compiler technology. Furthermore, RISC processors rely very much on auxiliary processor features–such as a large number of general-purpose registers, instruction and data caches, and so on–that help the compiler reduce the overall instruction count or that reduce the

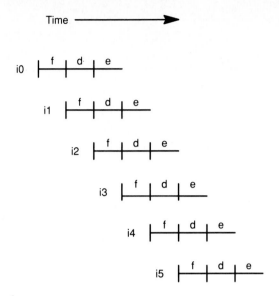

Figure 1-5. A RISC Processor Without Memory Limitations

number of cycles per instruction. These auxiliary features often require a significant amount of hardware.

The techniques we have examined in Figure 1-1 through Figure 1-5 have compressed the execution time up to an apparent barrier. The RISC processor is executing one instruction on every processor cycle, and no more improvement is possible given our current assumptions. However, there is no real reason to stop here. The barrier is due only to our assumption that each pipeline stage is one instruction wide and handles only one instruction per cycle. Figure 1-6 shows what happens if we relax this assumption, in a *superscalar* processor. We will use the term *scalar* processor to denote a processor that executes one instruction at a time, as in Figure 1-5. A superscalar processor reduces the average number of cycles per instruction beyond what is possible in a pipelined, scalar RISC processor by allowing concurrent execution of instructions *in the same* pipeline stage, as well as concurrent execution of instructions in different pipeline stages. The term superscalar emphasizes multiple, concurrent operations on scalar quantities, as distinguished from multiple, concurrent operations on vectors or arrays as is common in scientific computing.

Obviously, there is much more to the design of a superscalar processor than is implied by Figure 1-6, or there would not be much more to this book. As we will see, there are many things standing in the way of concurrent instruction execution, especially for general-purpose applications. Most of this book is not about making the

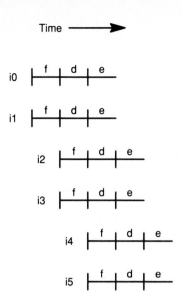

Figure 1-6. Instruction Timing in a Superscalar Processor

processor pipelines wider than one instruction, but rather about making this pay off. We will look at a wide range of limitations and trade-offs in the superscalar processor. We will explore and evaluate a number of processor features both in terms of their costs and in terms of the performance they provide, and we will investigate areas where performance is not very sensitive to hardware simplification. We will also see how software contributes to performance, see how the hardware both interferes with and compliments software, and see how to make trade-offs between software and hardware techniques.

Chapter 2

An Introduction to Superscalar Concepts

Superscalar processors are conceptually simple, but there is much more to achieving performance than widening a processor's pipeline. Widening the pipeline makes it possible to execute more than one instruction per cycle, but there is no guarantee that any given sequence of instructions can take advantage of this capability. Instructions are not independent of one another, but are interrelated; these interrelationships prevent some instructions from occupying the same pipeline stage. Furthermore, the processor's mechanisms for decoding and executing instructions can make a big difference in its ability to discover instructions that can be executed at the same time.

A major theme of this book is that superscalar processors are the natural next step in the evolution of general-purpose microprocessors, beyond RISC processors. Superscalar techniques largely concern the processor organization, independent of the instruction set and other architectural features. Thus, one of the attractions of superscalar techniques is the possibility of developing a processor that is code compatible with an existing architecture. Many superscalar techniques apply equally well either to RISC or CISC architectures. However, because the properties of a RISC processor make it easier to understand the application of superscalar techniques and thus simplifies our investigation, we will concentrate on RISC processors and mention topics relevant to CISC processors only in passing.

2.1 FUNDAMENTAL LIMITATIONS

A sequence of instructions can have three properties that limit the performance of a superscalar processor: *true data dependencies, procedural dependencies*, and *resource conflicts*. These limits are caused by fundamental characteristics of processors and set an upper bound on performance. (Some readers already familiar with this topic may notice that two limitations are missing. These will be introduced later in a more appropriate context.)

This section describes true data dependencies, procedural dependencies, and resource conflicts. It also explains how they limit performance. We will see how these limitations arise differently in RISC and CISC processors, to help motivate the focus on RISC processors throughout the rest of this book. Both RISC and CISC processors suffer the same types of limitations, but the limitations are more complex to understand for CISC processors and are more difficult for the processor to detect and deal with.

2.1.1 True Data Dependencies

Most processor instructions produce result values, and these values are used as input values to other instructions. If an instruction uses a value produced by a previous instruction, the second instruction has a *true data dependency* on the first instruction (also *true dependency*, *data dependency,* or, in this book, simply *dependency*). If the superscalar processor attempts to execute these two instructions at the same time, the execution of the second instruction must be delayed until the first instruction produces its value (see Figure 2-1). In general, any instruction must be delayed until all of its input values have been produced. This limits the number of instructions available to be executed at any given time.

True data dependencies also limit the performance of a scalar, pipelined processor. For example, a typical RISC processor takes two or more cycles to produce the result value of a *register-load* instruction, because the load involves the delay of an off-chip memory or cache. If the immediately following instruction uses the data produced by the load, instruction execution is suspended until the load completes. It is common for RISC compilers to schedule the execution of a load instruction so that one or more instructions—which do not depend on the load—are executed before an instruction that does depend on the load. Unfortunately, the independent instructions executed by a scalar processor during the load tend to be executed by a super-

Figure 2-1. Effect of True Data Dependencies on Instruction Timing

scalar processor on the first cycle of the load, leaving the superscalar processor with nothing to do while the load completes. The limitation of true dependencies is more severe in a superscalar processor than in a scalar processor, because it prevents a greater amount of processing from being accomplished.

The longer an operation takes to execute, the less likely it is that the processor will find other instructions to overlap with the long operation. At some point, it is likely there will be no instruction that is not dependent on the result of the long operation. This creates a number of *zero-issue* cycles, where the processor does not execute any new instruction (although it is still in the process of executing the long operation). When the long operation does complete, there is a reasonable chance that only one new instruction can be executed, because all subsequent instructions depend in turn on this new instruction. This creates a *single-issue* cycle, where the processor can execute only one instruction.

During zero-issue and single-issue cycles, the ability of the processor to execute more than one instruction at a time is irrelevant, and wide processor pipelines are wasted. Operation times, or *latencies*, have an important bearing on the effectiveness of the superscalar processor. For example, if operation latencies cause one-third of all cycles in a program to be single-issue cycles, wide pipelines cannot speed up the program by more than a factor of three, assuming that all other cycles can be reduced to near zero by a near-infinite pipeline width [Jouppi 1989b].

By nature, the compound, multicycle operations that characterize CISC processors have many true dependencies between the subfunctions of the instruction–in fact, many CISC instructions are inherently serial. For example, consider an old CISC standby–the *register-to-memory add* instruction. This instruction loads a value from a memory location, adds it to a value in a register, and stores the result back into the memory location. There are no parts of this instruction that can be executed at the same time. Each subfunction must wait for the previous subfunction to complete, so these long operations cause zero-issue and single-issue cycles and reduce the effectiveness of wide pipelines. For instructions such as these, the only opportunities for parallel execution are between operations in different instructions. However, complex instructions use a number of input values and produce a number of result values, making it difficult to determine the dependencies between instructions. Even worse, many times the input and output operands of complex instructions–for example, values in segment registers or special status registers–are *implied* by the instruction or by a mode of the instruction, rather than being explicitly called out by the instruction. This allows the instructions to be encoded more densely, but adds a lot of decoding overhead to the hardware that detects dependencies between instructions. In a RISC processor, this dependency logic need do little more than compare register identifiers that appear in fixed locations of the instructions.

Having complex instructions that take a long time to execute, with instruction dependencies that are difficult to detect, obscures the possibility of executing parts of the instructions at the same time. Breaking instructions into smaller components where each component explicitly calls out its input values and identifies its results values–as in a RISC processor–exposes more potential overlap between instructions. It is easier to overlap small, uniform operations than large, nonuniform ones.

2.1.2 Procedural Dependencies

If a processor knew, at the beginning of a program, exactly which instructions would be executed, very high performance would be possible [Riseman and Foster 1972, Nicolau and Fisher 1984]. Unfortunately, because of branches, the processor discovers the set of instructions to be executed only as execution progresses. The instructions following a given branch instruction have a *procedural dependency* on that branch and cannot be executed until the branch is executed. Because an instruction may be executed more than once, a given execution of an instruction has a procedural dependency on each branch that has been executed up to that point. Figure 2-2 shows how procedural dependencies limit superscalar performance: when the processor waits for a branch to complete so that it knows which instructions to execute next, it cannot execute subsequent instructions in parallel.

The same sort of branch limitation applies in a scalar, pipelined processor, because branches prevent the overlap of instructions in different pipeline stages [McFarling and Hennessy 1986]. In Figure 2-2, the delay between the execution of

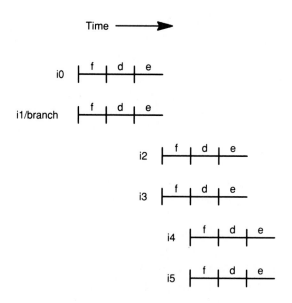

Figure 2-2. Effect of Procedural Dependencies on Instruction Timing

instructions *i1* and *i2* would exist even in a scalar processor. However, the delay is relatively more severe in a superscalar processor, because it prevents the execution of a potentially greater number of instructions while the processor determines the outcome of the branch.

CISC processors have another procedural dependency in addition to that caused by branch instructions. Because encoding efficiency is one of the objectives of a CISC instruction set, most (if not all) CISC processors have variable-length instructions; that is, instructions are represented by a variable number of bits so that the encoding of a given instruction can be made as small as possible. When an instruction is fetched, the processor does not know the location of any succeeding instruction until it determines the length of the fetched instruction. Usually, the length is determined during the decode stage. If the processor waits until after the decode stage before it knows the location of the next instruction, it cannot very well execute the next instruction at the same time. A CISC processor requires special techniques to be able to locate more than one instruction per cycle. In Chapter 4, where we consider procedural dependencies in detail, we will reap many advantages from assuming that all instructions have fixed length, a characteristic that applies to most RISC processors.

2.1.3 Resource Conflicts

An instruction uses processor resources at each stage of execution: these resources include memories, caches, decoders, buses, register-file ports, functional units, and so on. A *resource conflict* arises when two instructions must use the same resource at the same time. As Figure 2-3 shows, a resource conflict is resolved by delaying the execution of one of the conflicting instructions, so that the instructions use the resource at different times.

Resource conflicts also arise in scalar, pipelined processors. For example, an instruction fetch may contend with a data-load operation for a memory bus. However, a superscalar processor has a much larger number of potential conflicts, because instructions are attempting to occupy the same pipeline stage and, therefore, to

Figure 2-3. Effect of Resource Conflicts on Instruction Timing

use the resources dedicated to that pipeline stage. The average utilization of a resource in the scalar processor does not necessarily indicate the frequency of resource conflicts in the superscalar processor, because conflicts are determined by the timing of resource use. For example, a resource that is used by every other instruction in the scalar processor may, in a superscalar processor that attempts to execute two instructions per cycle, create either no resource conflicts or a conflict every third cycle, depending on which instructions the superscalar processor attempts to execute at the same time (the conflict occurs every three cycles, rather than every two cycles, because an additional cycle is needed to resolve the conflict).

In principle, it is easy to remove a resource conflict, but this is not necessarily cost effective. A conflict is eliminated simply by duplicating the troublesome resource. However, this can be very expensive, for example when the resource is a main memory port or a large functional unit such as a floating-point multiplier. Removing the resource conflict also may not improve performance very much because other limitations, such as true dependencies and procedural dependencies, are not removed by duplicating the resource and may very well prevent any performance improvement.

Figure 2-3 might imply that resource conflicts have an effect on performance that is similar to the effect of true dependencies (Figure 2-1). However, there are a few important differences. Resource conflicts can be eliminated by the duplication of resources, whereas true dependencies cannot be eliminated (except by reducing the execution time to zero, which is impossible). Also, when an operation takes a long time to complete, resource conflicts between two similar operations can be significantly reduced by pipelining the requisite functional unit, as shown in Figure 2-4. Pipelining is typically much less expensive than duplicating the functional unit. In contrast, there is nothing that can be done for an instruction that has a dependency on the result of a long operation.

In a sense, resource conflicts can be viewed as an indication of a benefit. Processors typically contain a number of independent functional units: an arithmetic-logic unit (ALU), a shifter, a branch unit, a load/store unit, a floating-point unit, and so on. In scalar processors, these independent units are used one at a time, because the processor can issue only one instruction at a time. A superscalar processor can use these functional units at the same time, and thus uses them more efficiently. The presence of a few conflicts is just a natural result of this efficiency, as long as resource conflicts do not severely limit performance.

The resource-utilization patterns of CISC processors make it generally more difficult to predict when resource conflicts will arise and to deal with them when they do arise. For example, consider again the *register-to-memory add* instruction. This instruction might use resources in the following sequence during its execution: the instruction memory, the instruction decoder, a register-file port, an ALU (for address computation), a memory address bus, a data memory, a memory data bus, a register-

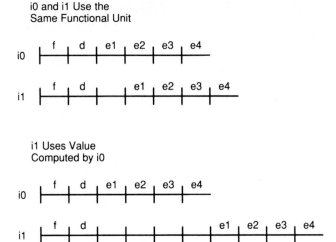

Figure 2-4. Resource Conflicts and True Dependencies for Long Operations

file port (to store intermediate data into a temporary register), two register-file ports (to read operands), an ALU (to perform the add), the memory-address bus, the memory-data bus, and the data memory. This presents no particular problem in a scalar processor, where the instruction assumes control of the processor at the beginning of its execution and maintains control until the end. In this case, there is no other instruction to cause conflicts. In contrast, a superscalar processor attempts to execute two or more instructions at the same time. Each instruction, on each cycle of its execution, must generate the controls that make the other instruction(s) aware of its intention to use shared resources. When instructions use many resources over many cycles, the potential conflicts between multiple instructions can be tremendously complex. The CISC processor requires very complex control hardware to sustain two or more arbitrary instructions in concurrent execution [Patt et al. 1985a/b, Hwu and Patt 1986].

2.1.4 Instruction Parallelism and Machine Parallelism

The *instruction parallelism* of a program is a measure of the average number of instructions that a superscalar processor might be able to execute at the same time–mostly, the instruction parallelism is determined by the number of true dependencies and the number of branches in relation to other instructions. Instruction parallelism also is determined by the processor, because the latencies of processor operations determine how severely true dependencies limit the number of instructions that can be executed at the same time.

The *machine parallelism* of a processor is a measure of the ability of the processor to take advantage of the instruction-level parallelism. Machine parallelism is determined by the number of instructions that can be fetched and executed at the same time and by the speed and sophistication of the mechanisms that the processor uses to find independent instructions.

To achieve performance, both machine parallelism and instruction parallelism are required [Jouppi 1989b]. Some programs do not have enough instruction parallelism to take advantage of machine parallelism, and some programs do not achieve their full performance potential because of limited machine parallelism. A challenge in the design of superscalar processor is to achieve a good balance between instruction parallelism and machine parallelism. Some programs have so much instruction parallelism that they justify large amounts of processor hardware. For example, consider the following code fragment:

```
DO I = 1 TO 100
    A[I] = A[I] + 1
CONTINUE
```

In this fragment, there are a large number of independent computations and the number of computations is known a priori. We could conceivably design a processor with 100 arithmetic/addressing units and 100 memory ports and achieve a speedup of 100 over a scalar processor. Of course this is a contrived example, but the point is that some programs–particularly those representative of scientific applications–have special characteristics that allow huge performance gains from machine parallelism. It is important not to focus on such programs during the design of the processor, because general-purpose programs often do not have such characteristics. If such programs really do represent the programs that the processor executes most of the time, there are many techniques other than superscalar techniques that might be more appropriate (see Section 2.4). As stated before, we will concentrate mostly on general-purpose applications in this book.

The question of whether RISC processors or CISC processors yield more theoretical parallelism is interesting, but it is a question we will ignore here. We will concentrate on RISC processors in this book, because our purpose is to develop insight into and understanding of superscalar processors. As we have seen in this section, CISC processors introduce several unique problems, and I believe that exploring solutions to these problems contributes little to understanding the fundamentals of superscalar design. Also, our approach is likely to be more relevant to the actual development of superscalar technology. RISC processors address the needs of performance-oriented users, and so are more likely to have impetus toward superscalar designs than CISC processors, for which code compatibility is more important than performance. Furthermore, having less processor circuitry and designers' mental energy devoted to complex instructions allows more to be devoted to parallelism, so that RISC processor can adopt superscalar techniques more easily than CISC processors.

However, the purpose of this book is *not* to serve as one more treatise in the CISC-versus-RISC debate. CISC processors can take advantage of the techniques we explore, but these will likely be applied to common, frequent instruction sequences. The most complex instructions can be executed in a slower, sequential mode to avoid the problems that are introduced by concurrent execution. This may place limits on the potential advantage of superscalar techniques in CISC processors, but it does not mean such techniques cannot be used. We will bring up special considerations related to CISC processors only to contribute to understanding or illustrate a pitfall to be avoided. For interested readers, the Appendix explores a superscalar design of what many regard as the ultimate successful CISC architecture–the Intel 386™.

2.2 INSTRUCTION ISSUE AND MACHINE PARALLELISM

The instruction parallelism of a program is not necessarily exploited simply by widening the processor's pipeline and adding hardware resources. The processor's policies toward fetching, decoding, and executing instructions have a significant effect on its ability to discover instructions which it can execute concurrently. We use the term *instruction issue* to refer to the process of initiating instruction execution in the processor's functional units and the term *instruction-issue policy* to refer to the protocol used to issue instructions.

The instruction-issue policy limits or enhances performance because it determines the processor's *lookahead* capability [Keller 1975]; that is, the ability of the processor to examine instructions beyond the current point of execution in hopes of finding independent instructions to execute. For example, if a resource conflict between two instruction halts instruction fetching, the processor is not able to look any further in the instruction sequence until after the conflict is resolved, because no more instructions are fetched. But, if the processor does continue to fetch instructions in spite of the conflict and, as a result, does find an independent instruction, the processor must be able to execute the independent instruction without regard for the conflicting instructions, if it is to take advantage of lookahead. The instruction-issue policy therefore affects lookahead by determining the order in which instructions are allowed to execute and complete. The processor can be more and more sophisticated, with better and better performance, by paying less and less attention to the relationship between the order in which instructions are fetched, the order in which they are executed, and the order in which they change processor memory locations (or *state*). However, the processor cannot look ahead arbitrarily, because it must behave correctly and produce correct results. To insure correctness, the more sophisticated an instruction-issue policy is, the more complex it is.

2.2.1 In-Order Issue with In-Order Completion

The simplest instruction-issue policy is to issue instructions in exact program order (*in-order issue*) and to write their results in the same order (*in-order completion*). Figure 2-5 illustrates the operation of a superscalar processor with in-order issue and in-order completion. To help visualize the operation of the superscalar processor, this diagram shows the processor pipeline stages horizontally and shows clock cycles vertically. This particular processor can decode two instructions, execute them in three functional units, and write two results per cycle (during the *writeback* stage of the pipeline). The instruction sequence shown in Figure 2-5 has the following constraints on parallelism:

- *I1* requires two cycles to execute.
- *I3* and *I4* conflict for a functional unit.
- *I5* depends on the value produced by *I4*.
- *I5* and *I6* conflict for a functional unit.

In this case, the pipeline is designed to handle a certain number of instructions (Figure 2-5 shows two instructions, because of decode and writeback limitations), and only this number of instructions can be in execution at once. Instruction results are written back in the same order that the corresponding instructions were fetched. To accomplish this, instruction issuing stalls when there is a conflict for a functional unit (the conflicting instructions are then issued in series) or when a functional unit requires more than one cycle to generate a result. For this example sequence, the total time from decoding the first instruction to writing the last results is eight cycles.

In-order issue with in-order completion is described here mainly to provide contrast with the more complex schemes described in the next few sections. In-order completion is not used even in scalar processors, because it causes long-latency operations to limit performance more than is justified by its simplicity.

Decode		Execute			Writeback		Cycle
I1	I2						1
I3	I4	I1	I2				2
I3	I4	I1					3
	I4			I3	I1	I2	4
I5	I6			I4			5
	I6		I5		I3	I4	6
			I6				7
					I5	I6	8

Figure 2-5. Superscalar Pipeline with In-Order Issue and In-Order Completion

2.2.2 In-Order Issue with Out-of-Order Completion

The next step in complexity from in-order completion is *out-of-order completion*. This technique is used in scalar RISC processors to improve the performance of long-latency operations such as loads [AMD 1989] or floating-point operations [Motorola 1989]. Out-of-order completion also improves the performance of a superscalar processor for the same types of operations. Figure 2-6 illustrates the operation of a superscalar processor with in-order issue and out-of-order completion, using the example of Figure 2-5.

With out-of-order completion, any number of instructions is allowed to be in execution in the functional units, up to the total number of pipeline stages in all functional units. Instructions may complete out of order because instruction issuing is not stalled when a functional unit takes more than one cycle to compute a result. Consequently, a functional unit may complete an instruction after subsequent instructions already have completed. In Figure 2-6, instruction *I1* completes out of order, improving the processor's lookahead ability: note that, in comparison to Figure 2-5, *I3* is executed concurrently with the last cycle of *I1*. The total time of this sequence has been reduced to seven cycles.

In a processor using out-of-order completion, instruction issuing is stalled when there is a conflict for a functional unit or when an issued instruction depends on a result that is not yet computed. And there is one more situation that stalls instruction issuing–one that we have not considered yet. To understand this, consider the following code sequence ("op" is any operation, "R*n*" represents a numbered register, and ":=" represents assignment):

```
R3 := R3 op R5      (1)
R4 := R3 + 1        (2)
R3 := R5 + 1        (3)
R7 := R3 op R4      (4)
```

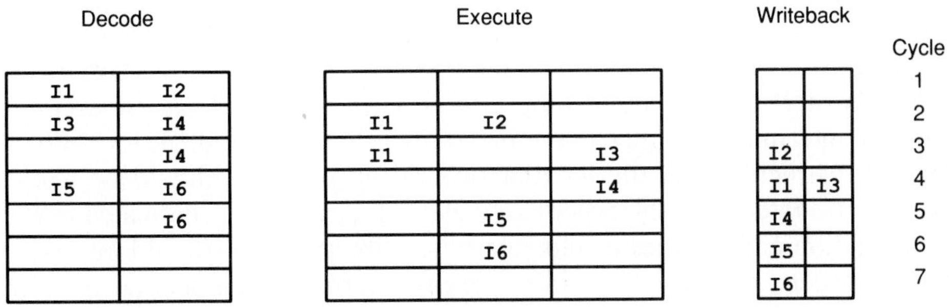

Decode			Execute			Writeback		Cycle
I1	I2							1
I3	I4		I1	I2				2
	I4		I1		I3	I2		3
I5	I6				I4	I1	I3	4
	I6			I5		I4		5
				I6		I5		6
						I6		7

Figure 2-6. Superscalar Pipeline with In-Order Issue and Out-of-Order Completion

In this example, the assignment of the first instruction cannot be completed after the assignment of the third instruction, even though instructions may in general complete out of order. Completing the first and third instructions out of order would leave an old, incorrect value in register *R3*, possibly causing, for example, the fourth instruction to receive an incorrect operand value. The result of the third instruction has an *output dependency* on the first instruction: the third instruction must complete after the first instruction to produce the correct output values of this code sequence. To insure this, the issuing of the third instruction must be stalled if its result might later be overwritten by an older instruction which takes longer to complete.

Out-of-order completion yields higher performance than in-order completion, but requires more hardware than in-order completion. Dependency logic is more complex with out-of-order completion, because this logic checks data dependencies between decoded instructions and all instructions in all pipeline stages. Hardware also must insure that results are written in a correct order. In contrast, with in-order completion, the dependency logic checks data dependencies between decoded instructions and the few instructions in execution, and results are naturally written in the correct order. Out-of-order completion also creates a need for functional units to arbitrate for result buses and register-file write ports, because there probably are not enough of these to satisfy all instructions that can complete simultaneously.

Out-of-order completion also makes it more difficult to deal with instruction exceptions. Often, an instruction creates an exception when, under the given conditions, the instruction cannot be properly executed by hardware alone. It is common to handle these exceptional conditions by causing an interrupt, permitting a software routine to correct the situation. Once the software routine has been completed, it is necessary to restart the execution of the interrupted program so that it can continue as usual. However, out-of-order completion complicates restarting, because the exceptional condition may have been detected as an instruction produced its result out of order. The program cannot be restarted at the instruction following the exceptional instruction, because subsequent instructions have already completed, and doing so would cause these instructions to be executed twice.

One approach to restarting after an exception relies on processor hardware to maintain a simple, well-defined restart state that is identical to the state of a processor having in-order completion [Smith and Pleszkun 1985]. A processor providing this form of restart state is said to support *precise interrupts* or *precise exceptions*. With precise interrupts, the interrupt return address indicates both the location of the exceptional instruction and the location where the program should be restarted. Without precise interrupts, the processor needs a mechanism to indicate the exceptional instruction and another to indicate where the program should be restarted. With out-of-order completion, providing precise interrupts is harder than not providing them, because of the hardware required to give the appearance of in-order completion.

2.2.3 Out-of-Order Issue with Out-of-Order Completion

With in-order issue, the processor stops decoding instructions whenever a decoded instruction creates a resource conflict or has a true dependency or an output dependency on a uncompleted instruction. As a result, the processor is not able to look ahead beyond the instructions with the conflict or dependency, even though one or more subsequent instructions might be executable. To overcome this limitation, the processor must isolate the decoder from the execution stage, so that it continues to decode instructions regardless of whether they can be executed immediately. This isolation is accomplished by a buffer between the decode and execute stages, called an *instruction window*.

To take advantage of lookahead, the processor decodes instructions and places them into the window as long as there is room in the window and, at the same time, examines instructions in the window to find instructions that can be executed (that is, instructions that do not have resource conflicts or dependencies). The instruction window serves as a pool of instructions, giving the processor a lookahead ability that is constrained only by the size of the window and the capability of the instruction-fetch unit. Instructions are issued from the window with little regard for their original program order, so this method issues instructions *out of order*. The only constraints on instruction issue are those required to insure that the program behaves correctly.

Figure 2-7 illustrates the operation of a superscalar pipeline with out-of-order issue, using the instruction sequence of Figure 2-5 and Figure 2-6. The instruction window is not an additional pipeline stage, but is shown in Figure 2-7 between the decode and execute stages for clarity. The fact that an instruction is in the window only implies that the processor has sufficient information about the instruction to know whether or not it can be issued. In Figure 2-7, because of the buffering provided by the instruction window, the decoder can operate at the maximum rate and complete decoding the instruction sequence well before it is executed. This allows

Decode		Window	Execute			Writeback		Cycle
I1	I2							1
I3	I4	I1, I2	I1	I2				2
I5	I6	I3, I4	I1		I3	I2		3
		I4, I5, I6		I6	I4	I1	I3	4
		I5		I5		I4	I6	5
						I5		6

Figure 2-7. Superscalar Pipeline with Out-of-Order Issue and Out-of-Order Completion

the processor to discover the independent instruction *I6,* which it executes out of order, concurrently with *I4.* The total time of this sequence has been reduced to six cycles.

With out-of-order issue, the issue constraints on any particular instruction are mostly the same as with in-order issue: an instruction is issued when it is free of resource conflicts and dependencies. Out-of-order issue simply gives the processor a larger set of instructions available for issue, improving its chances of finding instructions to execute concurrently. However, the capability to issue instructions out of order introduces an additional issue constraint, much as the capability to complete instructions out of order introduced the constraint of output dependencies. To understand this, consider once more the example code sequence we used previously to illustrate output dependencies:

```
R3  :=  R3  op  R5        (1)
R4  :=  R3  +  1          (2)
R3  :=  R5  +  1          (3)
R7  :=  R3  op  R4        (4)
```

In this sequence, the assignment of the third instruction cannot be completed until the second instruction begins execution. Otherwise, the third instruction might incorrectly overwrite the first operand of the second instruction. The result of the third instruction has an *antidependency* on the first input operand of the second instruction. The term antidependency refers to the fact that the constraint is similar to that of true dependencies, except reversed. Instead of the first instruction producing a value that the second uses, the second instruction produces a value that destroys a value that the first one uses. To prevent this, the processor must not issue the third instruction until after the second one begins–and, since the second instruction depends on the first, the third instruction also must wait for the first to complete–even though the third instruction is otherwise independent.

Antidependencies are mainly of concern when instructions can issue out of order, because it is mainly in this situation that an input operand of a stalled instruction can be destroyed by a subsequent instruction during normal operation. However, in scalar processors, instruction exceptions are sometimes handled by correcting the exceptional condition, then retrying the problematic instruction. If this instruction completed out of order, then it is possible that, when it is retried, its input operands have been overwritten by subsequent instructions. This problem cannot occur in a processor that supports precise interrupts. In other cases, the solution may require that the processor maintain copies of instruction operands to allow restart [Pleszkun et al. 1987].

2.2.4 Storage Conflicts and Register Renaming

Readers familiar with the subject of instruction dependencies may be aware that true dependencies (sometimes called *flow dependencies* or *write-read* dependencies) are often grouped with antidependencies (also called *read-write* dependencies) and output dependencies (also called *write-write* dependencies) into a single group of instruction dependencies. The reason for this grouping is that each of these dependencies manifests itself through the use of registers or other storage locations. However, it is important to distinguish true dependencies from the other two. True dependencies represent the flow of data and information through a program. Anti- and output dependencies arise because, at different points in time, registers or other storage locations hold different values for different computations.

When instructions are issued in order and complete in order, there is a one-to-one correspondence between registers and values: at any given point in execution, a register identifier precisely identifies the value contained in the corresponding register. When instructions are issued out of order and complete out of order, the correspondence between registers and values breaks down, and values conflict for registers. This problem is especially severe when the processor's compiler performs register allocation [Chaitin et al. 1981], because the goal of register allocation is to keep as many values in as few registers as possible. Keeping a large number of values in a small number of registers creates a large number of conflicts when the execution order is changed from the order assumed by the register allocator [Hwu and Chang 1988].

Anti- and output dependencies are more properly called *storage conflicts*, because the reuse of storage locations (including registers) causes instructions to interfere with one another even though the conflicting instructions are otherwise independent [Backus 1978]. As we have seen, storage conflicts constrain instruction issue and reduce performance. But storage conflicts, like other resource conflicts, can be reduced or eliminated by duplicating the troublesome resource–in this case storage locations.

The processor removes storage conflicts by providing additional registers that are used to reestablish the correspondence between registers and values. The additional registers are allocated dynamically by hardware, and the registers are associated with the values needed by the program using *register renaming* [Keller 1975]. To implement register renaming, the processor typically allocates a new register for every new value produced: that is, for every instruction that writes a register. An instruction identifying the original register–for the purpose of reading its value–obtains instead the value in the newly allocated register. Thus, the hardware renames the original register identifier in the instruction to identify the new register and the correct value. The same register identifier in several different instructions may access different hardware registers, depending on the locations of register references with respect to register assignments.

With renaming, the example instruction sequence of Sections 2.2.2 and 2.2.3 becomes:

$$
\begin{aligned}
R3_b &:= R3_a \ op \ R5_a & (1) \\
R4_b &:= R3_b + 1 & (2) \\
R3_c &:= R5_a + 1 & (3) \\
R7_b &:= R3_c \ op \ R4_b & (4)
\end{aligned}
$$

In this sequence, each assignment to a register creates a new *instance* of the register, denoted by an alphabetic subscript. The creation of a new instance for *R3* in the third instruction avoids the anti- and output dependencies on the second and first instructions, respectively, and yet does not interfere with correctly supplying an operand to the fourth instruction. The assignment to *R3* in the third instruction *supersedes* the assignment to *R3* in the first instruction, causing $R3_c$ to become the new *R3* seen by subsequent instructions until another instruction assigns a value to *R3*.

Hardware that performs renaming creates each new register instance and destroys the instance when its value is superseded and there are no outstanding references to the value. This removes anti- and output dependencies and allows more instruction parallelism. Registers are still reused, but reuse is in line with the requirements of parallel execution. This is particularly helpful with out-of-order issue, because storage conflicts introduce instruction-issue constraints that are not really necessary to produce correct results. For example, in the preceding instruction sequence, renaming allows the third instruction to be issued immediately, whereas, without renaming, the instruction must be delayed until the first instruction is complete and the second instruction is issued.

2.3 RELATED CONCEPTS: VLIW AND SUPERPIPELINED PROCESSORS

There are other ways, beside superscalar techniques, to take advantage of instruction parallelism and improve performance. As we saw in Chapter 1, the time taken by a processor to complete a program is determined by three factors:

- the number of instructions required to perform the application
- the average number of processor cycles required to execute an instruction
- the processor cycle time

Superscalar techniques take advantage of instruction parallelism to reduce the average number of cycles per instruction. *Very-long-instruction-word (VLIW)* processors take advantage of instruction parallelism to reduce the number of instructions. *Superpipelined* processors take advantage of instruction parallelism to reduce the cycle time. Because they also exploit instruction parallelism, VLIW and superpipelined processors share many characteristics with superscalar processors and sometimes share common techniques for exploiting instruction parallelism. The distinction between these three architectural ideas is sometimes blurred. We will draw clearer distinctions here, limiting the scope of superscalar processors by defining what they are *not*.

2.3.1 Very-Long-Instruction-Word Processors

In a VLIW processor, a single instruction specifies more than one concurrent operation (Figure 2-8 shows an example of a processor which issues two operations using one instruction). Because a single VLIW instruction can specify multiple operations, in lieu of multiple scalar instructions, VLIW processors reduce the number of instructions for a program in comparison to scalar processors. However, in order for the VLIW processor to sustain an average number of cycles per instruction comparable to that of the scalar processor, the operations specified by the VLIW instruction must be independent of one another. Otherwise, the VLIW instruction is similar to a sequential, multiple-operation CISC instruction, and the number of cycles per instruction goes up accordingly.

As the name implies, the instruction of a VLIW processor is normally quite large, taking many bits to encode multiple operations. VLIW processors rely on software to pack the collection of operations representing a program into instructions. To accomplish this, software uses a technique called *compaction* [Fisher 1981, 1983]. The more densely the operations can be compacted (that is, the fewer the number of instructions used for a given set of operations), the better the performance, and the better the encoding efficiency. During compaction, *no-ops* (null operation fields) are used for instruction operations that cannot be used. In essence, compaction serves as a limited form of out-of-order issue, because operations can be placed into instructions in many different orders. To compact instructions, software

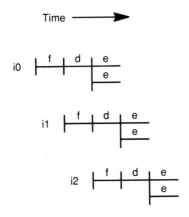

Figure 2-8. Instruction Timing in a VLIW Processor

must be able to detect independent operations, and this can restrict the processor architecture, the application, or both [Colwell et al. 1987].

VLIW processors have a few problems. First, and most important, they are not software compatible with any general-purpose processor. In fact, since the instruction parallelism–and therefore the compaction and the code for an application–depend on the processor's operation latencies, it is difficult to make different implementations of the same VLIW architecture binary-code compatible with one another. Second, the density of instruction compaction depends on the instruction parallelism. In sections of code having limited instruction parallelism, compaction is not dense, and most of the instruction is wasted. This can cause severe growth in the code size of a program–growth which must be addressed by special-purpose techniques such as instruction compression [Colwell et al. 1987] or by special serial instruction encodings [Cohn et al. 1989].

We will not dismiss VLIW techniques out of hand, however, because they lead to simple hardware implementations. A VLIW processor does not need hardware to detect instruction parallelism, because the parallelism is inherently specified by the instruction. In fact, we will use the term VLIW in general to refer to a processor that explicitly encodes parallelism in the instruction. We will defer examining VLIW processors until after we have had a chance to explore superscalar techniques in much more detail, because this makes it easier to place VLIW processors in perspective.

2.3.2 Superpipelined Processors

In a superpipelined processor, the major stages of a pipelined processor are divided into substages [Jouppi and Wall 1988]. The *degree of superpipelining* is a measure of the number of substages in a major pipeline stage. Figure 2-9 shows an example where two substages make up a major pipeline stage; the degree of superpipelining for this processor is two. The substages are clocked at a higher frequency than the major stage (the frequency is a multiple determined by the degree of superpipelining), and, using pipelining, the processor can initiate an operation at each substage on each of the smaller clock periods. This effectively reduces the processor cycle time, except that it relies on instruction parallelism to prevent pipeline stalls in the substages. For example, in Figure 2-9, if *i1* depends on the result of *i0*, its execution must be delayed for an additional cycle over that shown in the figure.

Comparing the operation of the superscalar processor in Figure 1-6 on page 7 with the operation of the superpipelined processor in Figure 2-9 illustrates several interesting differences between the two [Jouppi and Wall 1988, Jouppi 1989b]. For a given set of operations, the superpipelined processor takes longer to generate all results than the superscalar processor, and this can degrade performance in the presence of true dependencies. For example, two results that are available at the end of one cycle in the superscalar processor are not completely available until after the

Time ⟶

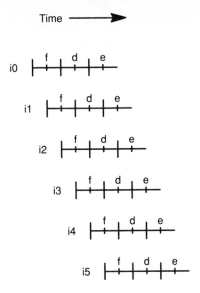

Figure 2-9. Instruction Timing in a Superpipelined Processor

equivalent of one and one-half cycles in the superpipelined processor. On the other hand, some simple operations in the superscalar processor take a full cycle only because there are no clocks with finer resolution. The superpipelined processor can complete these operations sooner. For example, when the outcome of a branch is determined by a simple zero-detect or equal-to operation, the superpipelined processor knows the outcome of a branch sooner than the superscalar processor, possibly reducing the impact of procedural dependencies.

There are more subtle differences between superscalar and superpipelined processors than those illustrated by Figure 1-6 and Figure 2-9. At a constant hardware cost, the superscalar processor is more susceptible to resource conflicts than the superpipelined processor. A resource must be duplicated to avoid conflicts in the superscalar processor, whereas the superpipelined processor avoids conflicts altogether, through pipelining (see Figure 2-4). However, superpipelining requires latches between substages, and these latches add somewhat to the overall cycle time. The higher the degree of superpipelining, the more severe the overhead of the pipeline latches.

The degree of superpipelining also impacts performance in another way. Part of the processor cycle is consumed by *clock skew*, or the relative timing of the physical clock signals that control various portions of the processor. For example, if the clocks controlling two sections of logic are skewed by 2 nanoseconds (ns), then, for these sections to communicate properly, the time allowed for communication must be the total delay of the communication, plus 2 ns. In a processor whose basic logic delays allow operation at 50 megaHertz (MHz), a clock skew of 2 ns represents 10% of the cycle time: the cycle time (20 ns) must increase by 10% to accommodate skew. In a comparable superpipelined processor, because the skew appears on clocks with a finer granularity, the skew represents a larger proportion of the cycle time, and the performance penalty caused by skew is relatively more severe by a factor determined by the degree of superpipelining. For example, in a comparable, 100 MHz superpipelined processor with superpipelining of degree two, a skew of 2 ns represents 20% of the minor cycle time: the minor cycle time and thus the major cycle time (again, 20 ns) must increase by 20% to accommodate skew. Alternatively, to avoid the superpipelined processor being slower than the superscalar processor, the technology of the superpipelined processor must reduce the clock skew to 1 ns for performance comparable to that of the superscalar processor. The superpipelined processor depends heavily on the ability of the implementation technology to generate clock signals with little skew.

To summarize these observations, superpipelining is an appropriate technique when the cost of duplicating resources is high and the ability to control skew is good. Thus, superpipelining is appropriate for processors implemented in very-high-speed technologies, such as bipolar, emitter-coupled logic or gallium arsenide. These technologies are characterized by low logic densities (so that duplicating resources is expensive) and very low gate delays (so that clock skew is well controlled). It is arguable whether such processors fit the category of "general-purpose microprocessors." However, our reasons for ignoring superpipelined processors in this book is that they present no new design problems over pipelined processors, which have already been extensively studied [Kogge 1981].

2.3.3 Hybrid Techniques

For the ambitious, it should be mentioned that superscalar, VLIW, and superpipelining techniques are not mutually exclusive. One could envision a superpipelined, superscalar, VLIW processor. However, the ultimate performance of this processor still is determined by the amount of instruction parallelism. If each technique were designed to allow two concurrent operations, then the three techniques combined would allow as many as eight concurrent operations. As we will see, few general-purpose applications have enough instruction parallelism to justify more than one technique, much less three.

2.4 UNRELATED PARALLEL SCHEMES

Most readers are probably aware of other types of parallel processors, of which there are many examples: vector processors, shared-bus multiprocessors, single-instruction/multiple-data (SIMD) processors, systolic arrays, hypercubes, dataflow architectures, and so on. We will say little about these other parallel schemes, except to point out an important difference between them and the three techniques discussed in Section 2.3, particularly superscalar processors. These other parallel schemes rely on a special property of some applications, called *data parallelism*. Data parallelism refers to the property that little, if any, two-way communication is required between different parts of a computation, or that, if two-way communication is required, the data in one direction does not depend on data in the other direction. Consider an example code sequence similar to one that we have already seen:

```
DO I = 1 TO 100
    B[I] = x * B[I]
    A[I] = B[I] + 1
CONTINUE
```

This sequence has a large amount of data parallelism, owing to the independent elements of the vectors *A[I]* and *B[I]*. Data parallelism makes it relatively inexpensive to perform the separate parts of a computation in independent computing elements. Because of data parallelism, these computing elements do not have to communicate very much, or communicate in a very systematic, predetermined fashion. For example, in the sequence given, separate elements might compute individual values *B[I]* and transmit these values to other elements computing *A[I]*. The processor can implement a straightforward, relatively inexpensive mechanism for communicating data between computations without reducing performance very much. With data parallelism, a large number of processor organizations is possible: witness the number of parallel machines listed earlier. These machines differ mainly in the style of communication needed for computation, but all rely on data parallelism.

In contrast, superscalar processors (and VLIW and superpipelined processors) rely on instruction parallelism, and in fact communicate data quite efficiently at random between parallel computations. Efficient communication is expensive when there is no fixed communication pattern, requiring multiple register ports and complex buses between functional units. However, these are required to improve the performance of applications having little data parallelism. A superscalar processor can certainly exploit data parallelism, because independent data implies independent instructions. However, the superscalar processor is not likely to be as efficient in exploiting data parallelism as other parallel schemes, because instruction parallelism does not in general yield many independent computations and thus does not motivate having the large number of computing elements (that is, functional units) required to exploit data parallelism fully.

Superscalar processors (and related techniques for exploiting instruction parallelism) have been proposed as a means of implementing processors that are more balanced than other parallel machines such as vector processors [Agerwala and Cocke 1987]. For example, vector processors have very good performance for applications having a lot of data parallelism (that is, vectorizable applications) and have relatively poor performance otherwise because they do not have the appropriate data-communication and control mechanisms to support instruction parallelism. Superscalar processors have lower performance for vectorizable applications because they do not support as many concurrent operations, but do not suffer as much otherwise because they can exploit instruction parallelism. Consequently, the performance of a superscalar processor is more predictable across a range of applications than is the performance of a vector processor.

Except for this observation, however, it is probably inappropriate to consider applications with large amounts of data parallelism in the design of a superscalar processor. There are many machine organizations that exploit data parallelism more effectively. Superscalar processors are more appropriate for applications having little data parallelism. Attempting to address both types of applications in the superscalar processor leads to more functional units than are required to exploit instruction parallelism, with more exotic interconnection than is required to exploit data parallelism. The resulting processor may be too expensive for applications of either type, compared to other alternatives. In the next chapter, we will explore the design of a superscalar processor using mostly general-purpose applications to avoid this difficulty.

Chapter 3

Developing an Execution Model

There is much published literature on superscalar processors and related subjects such as out-of-order issue and register renaming (Section 3.6 contains a brief overview of this literature). The published literature is a good source of ideas, but does not provide much information on which to base implementation decisions. Published evaluations concentrate on specific techniques and evaluate implementation trade-offs only within the context of a given technique. In this book, we will explore a wide range of design alternatives and investigate at length the feasibility of superscalar processors in general-purpose applications. As a result, the scope of this book is much broader than most of the published literature, and the viewpoint is quite different. This chapter explains how we will proceed in the following chapters.

Our investigation has several important elements. First, to evaluate a large number of alternatives, our methodology must permit us to explore a very large set of designs. Moreover, to provide a good foundation for our evaluation, the performance of the various superscalar designs should be measured against the performance of a scalar processor whose characteristics are well understood. Obviously, we must rely on simulation rather than measuring real machines. However, the simulation must be at a level of abstraction that is high enough to permit exhaustive investigation but low enough to reflect detailed design decisions. The evaluation of trade-offs must have the nature of experimentation. We cannot simply look at a lot of processor configurations, but must gauge the relative benefits or disadvantages of the alternatives.

Second, our evaluation should be based on a large set of programs in everyday use. By using large, general-purpose programs, we are more likely to avoid using programs with special characteristics that exaggerate the performance of the superscalar processor. In a similar vein, these programs should be generated with a highly optimizing compiler. Unoptimized code typically has more instruction parallelism than optimized code, because it contains unneeded and redundant instructions. A superscalar processor can execute this unnecessary code more efficiently than a scalar processor, because it can overlap unnecessary instructions with important ones. The scalar processor must be put on the best footing to avoid overstating the performance advantage of the superscalar processor.

Finally, we should measure, to the extent allowed by simulation, the performance of real processors in real systems. Theoretical architectures have their place, but I believe that readers' interests are best served by focusing on well-understood processor and system configurations. Our measurements are of little use if they ignore important system effects such as cache misses.

3.1 SIMULATION TECHNIQUE

We will use *trace-driven simulation* to evaluate the benefits of various superscalar organizations. Trace-driven simulation uses a predetermined instruction sequence, an *instruction trace*, to evaluate superscalar performance. In our case, the instruction trace is the sequence of instructions in a program as they are executed by a scalar processor (this scalar processor also is used as a basis for comparing the relative performance of the superscalar processor). Trace-driven simulation is efficient, because the simulator is concerned only with the processor features that affect performance. For example, the simulator need not track the contents of every processor register: it is only necessary that the simulator be aware of register identifiers in the instructions. Furthermore, the simulator need not execute instructions or model the detailed operation of the functional units. To determine performance, the simulator simply models the functional units as elements which delay the operand values needed by some instructions and which also prevent the simultaneous execution of some instructions.

The simulator is based on the MIPS, Inc., R2000™ RISC processor and R2010™ floating-point unit [Kane 1987]. The R2000 has very good optimizing compilers, allowing us to meet the objective of using highly optimized code. Also, the performance of this architecture is well characterized, and it serves as a good standard for measuring performance [MIPS 1989, Waterside 1989].

In order to limit the scope of the trade-offs we investigate, we will not vary the fundamental characteristics of the processor. Instead, we will model the characteristics of the superscalar processor after the characteristics of the R2000 processor and R2010 floating-point unit; these are fairly representative implementations. The R2000 and R2010 have the functional units shown in Table 3-1. Table 3-1 also gives the latencies of these functional units. The *issue latency* is defined as the minimum number of cycles between the issuing of an instruction to a functional unit and

Table 3-1. Configuration of Functional Units

Functional Unit	Issue Latency (cycles)		Result Latency (cycles)	
	single	double	single	double
Integer ALU (2)	1	n/a	1	n/a
Barrel Shifter	1	n/a	1	n/a
Branch Unit	1	n/a	1	n/a
Load Unit	1	2	2	3
Store Unit	1	2	n/a	n/a
Memory (cache reload)	13	n/a	12	n/a
Float Add	1	2	2	3
Float Multiply	1	2	4	6
Float Divide	12	27	12	27
Float Convert	1	2	2	4

the issuing of the next instruction to the same functional unit. The *result latency* is the number of cycles taken by the functional unit to produce a result. Issue and result latencies can depend on whether an operand is a single (32-bit) word or a double-precision (64-bit) floating-point number.

The scalar and superscalar processors have identical caches and memory interfaces. The load/store latencies given in Table 3-1 are achieved using a 64-kilobyte, direct-mapped data cache with a four-word block size [Smith 1982]. A similar cache is used to supply instructions. The instruction and data caches are loaded via a 64-bit interface to the main memory. Upon a cache miss, the main memory can begin reloading the cache after an initial access time of 12 cycles (this access time reflects a high anticipated processor operating frequency). Once the reload begins, it is completed in 2 cycles (reflecting an interleaved memory design). Thirteen cycles are taken from the time a miss occurs until the next miss can be handled.

The instruction traces are generated as diagrammed in Figure 3-1. The optimized object code for a program is processed by *pixie*™ [MIPS 1986], a program that annotates the object code at the target of every branch and at each memory reference. The annotated code can be executed as the original program and has the same behavior as the original program. However, when the annotated code is executed, the instructions added by *pixie* output a dynamic trace stream in addition to the program's normal output. This trace stream consists of branch target addresses, counts of the number of instructions up to the next branch, and the addresses referenced by loads and stores. The simulator takes the trace stream as input and, using the original ob-

---

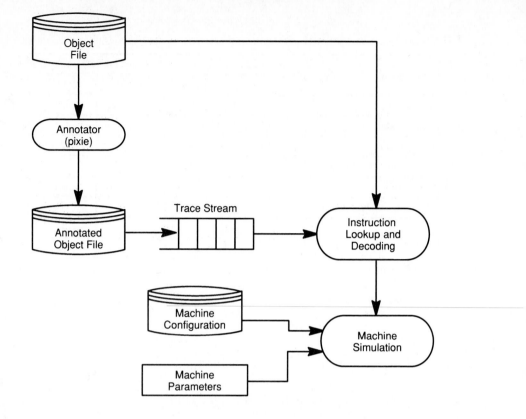

Figure 3-1. Flowchart for Trace-Driven Simulation

ject file, generates the full instruction trace needed by the simulator (this is required because the trace stream gives only branch target addresses and the number of instructions between branches).

The machine simulator models the functionality of both a scalar and a superscalar processor, using machine parameters specified when the simulator is invoked. The execution of the scalar and superscalar processors is modeled at a functional level, with time recorded in terms of machine cycles. The cycle time and functional-unit characteristics of the two processors are identical, although the superscalar processor can have more functional units and other resources.

The models of the scalar and superscalar processors differ primarily in the way they process instructions. The scalar processor issues instructions in order and completes instructions out of order, permitting a limited amount of concurrency between instructions (for example, between loads or floating-point operations and other integer operations). The superscalar processor is able to decode and issue any number of instructions, using any issue policy (the actual processor configuration is specified when the simulator is invoked). By keeping the number of instructions completed by each processor equal, the simulator is able to determine the performance advantage of the superscalar processor over the scalar processor in a way that is independent of other effects.

Except for the fact that the simulator does not actually execute instructions, its operation closely parallels the operation of the scalar and superscalar processors. As the simulator fetches instructions from the dynamic trace stream, it accounts for penalties which would be associated with fetching these instructions in an actual implementation (such as instruction-cache misses) and supplies these instructions to the scalar and superscalar processor models. Each processor model accounts for penalties caused by branches and by the execution unit not being able to accept decoded instructions (because of resource conflicts, for example). Instructions are checked for dependencies on all other uncompleted instructions. When an instruction is issued, it flows through a simulated functional-unit pipeline. The instruction cannot complete until it successfully arbitrates for one of (possibly) several result buses. When the instruction does complete, it may resolve dependencies and release other instructions for issue.

3.2   BENCHMARKING PERFORMANCE

Programs vary widely in the amount of instruction parallelism, so it is possible to choose a set of programs that overstates superscalar performance. To avoid this, we will use benchmark programs which mostly represent a wide class of general-purpose applications. Table 3-2 describes each benchmark program. The benchmarks include a number of integer programs (such as compilers and text formatters) and scalar floating-point applications (such as circuit and timing simulators). Linpack and Whetstone are notable exceptions to the "general-purpose" classification–they are included because they are widely used benchmarks and they contain a large number of floating-point operations (which reduce instruction parallelism, as we will see).

In general, the benchmark applications are thought to have little instruction parallelism because of their complex control flow, and they form a good test of the generality of the superscalar processor. To avoid improving the instruction parallelism because of the lack of compiler optimization, highly optimized versions of these programs are used. No software reorganization is performed (see Chapters 11 and 12), except for usual RISC pipeline scheduling already existing in the code [Hennessy and Gross 1983].

To obtain processor measurements, each program is allowed to complete 4 million instructions. This reduces simulation time, and includes a fixed overhead for loading the caches in every measurement. This approach roughly approximates per-

Table 3-2. Benchmark Program Descriptions

Program	Description
5diff	text file comparison
awk	pattern scanning and processing
ccom	optimizing C compiler
compress	file compression using Lempel–Ziv encoding
doduc	Monte-Carlo simulation, double-precision floating-point
espresso	logic minimization
gnuchess	computer chess program
grep	reports occurrences of a string in one or more text files
irsim	delay simulator for VLSI layouts
latex	document preparation system
linpack	linear-equation solver, double-precision floating-point
nroff	text formatter for a typewriter-like device
simple	hydrodynamics code
spice2g6	circuit simulator
troff	text formatter for typesetting device
wolf	standard-cell placement using simulated annealing
whetstone	standard floating-point benchmark, double-precision floating-point
yacc	compiles a context-free grammar into LR(1) parser tables

formance in a multitasking environment, because, in such environments, programs are allowed to execute for a fixed interval before being suspended, and resume execution without instructions and data residing in the caches. Our measurement interval corresponds to about 100-300 task switches per second, depending on the cycle time and execution rate of a particular configuration. In practice, the performance does not change much after the first million instructions.

The *speedup* of the superscalar processor for a particular program is the number of cycles taken by the scalar processor to execute the program divided by the number of cycles taken by the superscalar processor to execute the same program. For a given processor configuration, the measurements are presented as high, average, and low speedups over all benchmarks. To compute average performance, we will use the harmonic mean. For a set of N programs, each with speedup s_n, the harmonic mean of all speedups is:

$$\text{harmonic mean} = N \left(\Sigma_N \, 1/s_n \right)^{-1}$$

The harmonic mean has the advantage that it assigns larger weighting to the programs with the smallest speedups. This helps reflect the impact that the speedups have on the final, overall execution time: the programs that improve the least represent the greatest proportion of the final execution time. For example, consider a set of two programs with about the same execution time in the scalar processor (which roughly applies in our case, because the programs execute for a fixed number of instructions). If the first program experiences a speedup of 2 and the second a speedup of 25, the average speedup computed using the harmonic mean is 3.7. The arithmetic mean is 13.5, and the geometric mean is 7.07. The harmonic mean is much smaller because it reflects the fact that the execution time of one program has been reduced by half, taking a time which is one quarter the original total execution time of both benchmarks. The execution time of the other benchmark has been reduced to almost nothing, in relative terms, and the total time is still about one quarter of the original time (actually 1/3.7, because the second benchmark still contributes something). The speedup is about four (actually 3.7). The arithmetic and geometric means do not accurately reflect the final execution time.

We will use four programs in Table 3-2, *ccom, irsim, troff,* and *yacc,* as a representative sample for demonstrating and clarifying certain points in the next several chapters. Using this small sample helps limit the amount of data presented for the purpose of illustration. However, we will draw no conclusion based on this small sample.

3.3 BASIC OBSERVATIONS ON HARDWARE DESIGN

Superscalar processors present many more design decisions and trade-offs than scalar processors. Superscalar performance is affected by all of the things that affect scalar performance–the latencies of functional units, the pipeline organization, cache organization, memory performance, and so on–but superscalar processors also introduce additional degrees of freedom. In the superscalar processor, we deal with decisions such as the peak number of instruction issued, the issue policy, the number of functional units and their interconnection, whether or not to implement register renaming, and so on. In this book, we are going to ignore the things that affect scalar performance, taking the R2000 and R2010 as good representative scalar designs, and will focus on the things that affect superscalar performance. Otherwise, there are far too many things to look at. This book would not serve its purpose if it were to degenerate into a study of caches, floating-point units, and the like.

Even by limiting the field of our investigation in this way, there are still all sorts of comparisons we could do. To compare design alternatives, we will focus on the speedup resulting from different design alternatives and introduce additional data only when it helps illustrate the phenomena underlying the performance. Furthermore, we will compare the performance of different design alternatives against a single, standard superscalar processor to avoid getting lost in a maze of different machines. This is a straightforward experimental technique, but it presents a problem: there is no existing standard processor, so we must develop one. However, to develop one, we must rely on design techniques that we are trying to evaluate in the first place.

To break this circularity, we will spend this section and the next developing a standard superscalar processor against which to measure other alternatives. This section explores the design in gross terms, and the next section refines the specific machine design.

3.3.1 The Philosophy of the Standard Processor

A superscalar processor used as a standard for comparison should be the highest-performance processor that we could reasonably conceive of implementing. If the standard processor has low performance, we may find that very few performance improvements make sense, because other limitations prevent an improvement from achieving its full potential. On the other hand, if we choose an "ideal" processor, such as one with an infinite number of single-cycle execution units, we may find that every hardware simplification looks bad next to this ideal. The performance of the standard processor should be the best we can realistically achieve. It is acceptable to go a little overboard–our goal, after all, is to explore cost-performance trade-offs rather than to claim a single design is the best. However, we should be able to imagine that the standard processor could be built.

We already have a good idea of what techniques allow the superscalar processor to achieve best performance: out-of-order issue of multiple instructions with register renaming. However, we have no idea of what performance to expect from these techniques. Even though performance is "best," it may not be "best" by enough to be worth the bother. Nor do we know to what extent we should apply these techniques. However, we have at our disposal instruction traces of many real applications. These traces tell us exactly which instructions are executed and in what order they are executed. We can analyze the traces to determine the performance we should expect of the superscalar processor and the rough characteristics that the processor should have. We run the risk of fooling ourselves because we ignore procedural dependencies and the problems of fetching instructions in an actual implementation, but we will return to this problem later.

3.3.2 Instruction Parallelism of the Benchmarks

The performance of the superscalar processor cannot exceed the performance allowed by the instruction parallelism of the benchmark programs. As we make the processor more and more capable, we should expect a point of decreasing returns because performance will be more and more limited by the instruction parallelism.

Hence, it is important to get a feel for the amount of instruction parallelism in the benchmark programs, so that we know where this limit is.

To get the best understanding of the limits of instruction parallelism, we should consider instruction parallelism limited only by true dependencies. We can remove resource conflicts by assuming infinite resources, can remove storage conflicts by renaming, and can remove procedural dependencies by analyzing traces in which the outcome of all branches is known. True dependencies, however, cannot be removed even in an ideal case such as this, and thus place the ultimate limits on instruction parallelism.

As we saw in Section 2.1.1, true dependencies give rise to zero-issue and single-issue cycles that cannot be improved by machine parallelism. The number of zero-issue and single-issue cycles is determined by the latencies of the processor's functional units. Figure 3-2 shows the total zero-issue and single-issue cycles for the benchmark programs, as a percentage of total cycles, assuming the latencies given in Table 3-1 and assuming infinite processor resources. Resource conflicts, storage conflicts, and procedural dependencies are ignored, but latencies of cache reload are included since they affect instruction parallelism. Note that the benchmarks with the most floating-point operations–*doduc*, *linpack*, and *whetstone*–have the largest

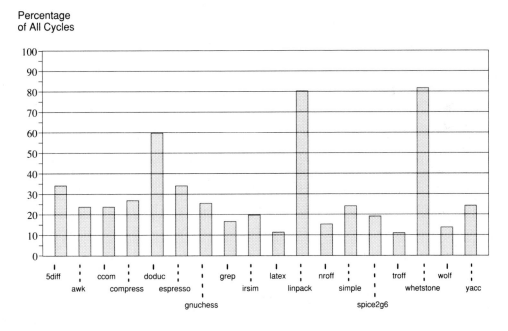

Figure 3-2. Percentage of Zero-Issue and Single-Issue Cycles in an Ideal Superscalar Processor with Given Latencies and Caches

number of zero- and single-issue cycles because of the longer latency of floating-point operations.

If we assume that, with infinite machine parallelism, all instructions can be executed in the time taken for the zero-issue and single-issue cycles, we obtain the maximum instruction-execution rate of these programs, shown in Figure 3-3. The average performance (the harmonic mean) is 3.3 instructions per cycle, even under the wholly unrealistic assumption that machine parallelism reduces to zero the number of cycles taken to complete the parallel-issue instructions. Thus, we are far from having to worry about issuing massive numbers of instructions in parallel.

Figure 3-3 shows the maximum instruction-issue rate of the superscalar processor, but we are more interested in the maximum speedup over the scalar processor. Determining the speedup requires that we know the instruction-execution rates of the scalar processor for the same programs; these are shown in Figure 3-4. The speedups are calculated by dividing the superscalar execution rate by the scalar execution rate and are shown in Figure 3-5. The harmonic mean of all speedups is 5.4. This is the highest average speedup that can be achieved, even with ideal machine parallelism. Some readers may be aware of claims of higher performance. It is important to realize that our results are for general-purpose benchmarks and that they include the latencies of cache misses. Many other claims that I am aware of are based

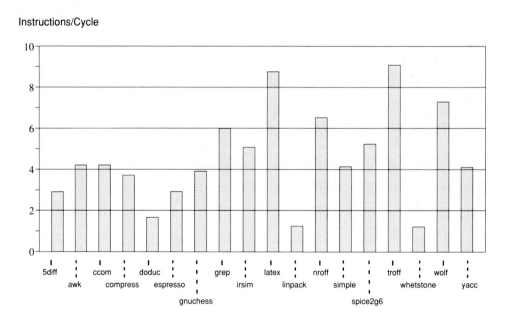

Figure 3-3. Maximum Instruction-Execution Rates of a Superscalar Processor with Given Latencies and Caches

Instructions/Cycle

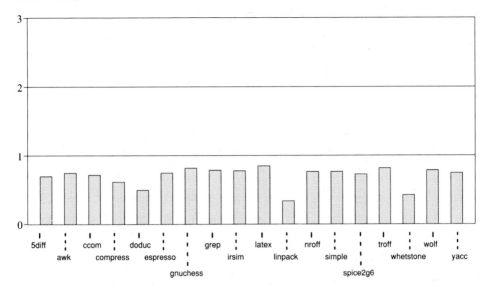

Figure 3-4. Instruction-Execution Rates of a Scalar Processor
with Given Latencies and Caches

on programs with large amounts of data parallelism and include further idealizations, such as single-cycle functional units and perfect, infinite caches, in addition to the idealizations we are using here.

The instruction parallelism places bounds on our quest for performance. By knowing the best performance we could achieve, we gain a better understanding of when to stop trying to improve performance.

3.3.3 Machine Parallelism

Having established the maximal level of instruction parallelism in the benchmarks, we now turn to the problem of a machine organization that exploits this parallelism. In the absence of procedural dependencies, we have at our disposal three techniques to improve machine parallelism: duplication of resources, out-of-order issue, and register renaming. In this section, we will establish the benefit of these techniques and determine to what extent they should be applied.

To explore the limits of machine parallelism, we will ignore long-latency operations. We do this by omitting the floating-point-intensive benchmarks and by assuming perfect, infinite caches that are preloaded with instructions and data. This is analogous to assuming infinite machine parallelism, in the previous section, in order to determine the limits of instruction parallelism. Here, we assume the best instruc-

---

Speedup

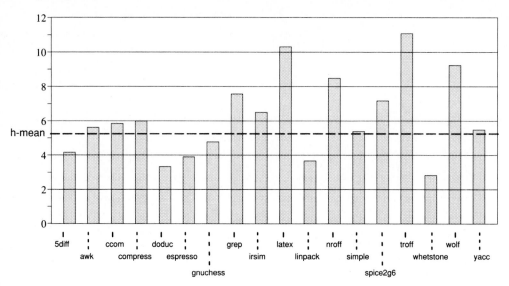

Figure 3-5. Maximum Speedups of a Superscalar Processor
with Given Latencies and Caches

tion parallelism we can imagine, in order to determine the limits of machine parallelism. Our intent is not to be overly optimistic, but to emphasize the existence of limits even under optimistic assumptions.

Figure 3-6 shows the average speedups of the benchmarks for several machine configurations, omitting the floating-point-intensive benchmarks and assuming that there are no cache misses. Out-of-order issue is assumed in both charts, but the processor's lookahead ability is varied by increasing the size of the instruction window. The chart on the left is the speedup without register renaming, and the chart on the right is the speedup with register renaming. Along the bottom axis are several machine configurations:

- base – the base processor, having the same number of functional units as the scalar processor. This superscalar processor has no more machine parallelism than the scalar processor, except for its ability to issue multiple instructions out of order with register renaming.

- +ld/st – a processor having an additional load/store unit, for a total of two load/store units. In one cycle, this processor can issue two loads, two stores, or one load and one store.

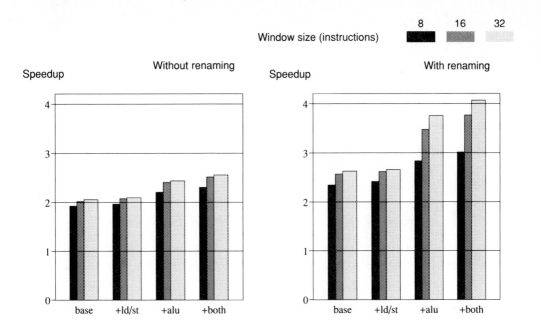

Figure 3-6. Speedups of Various Machine Organizations, Without Procedural Dependencies

- +alu – a processor having an additional arithmetic/logic unit, for a total of two ALUs. This processor can issue two arithmetic or logical instructions in one cycle.

- +both – a processor having an additional arithmetic/logic unit and load/store unit. This processor can issue two arithmetic or logical instructions and two load/stores in one cycle.

Figure 3-6 makes several things apparent. First, without register renaming, there is not much point in duplicating functional units. Second, with register renaming, we obtain the most benefit by duplicating the ALU. The benefit of duplicating the ALU is to be expected. By far the most frequent instructions in the MIPS architecture (and most other RISC architectures) are integer ALU instructions: these represent about 40-50% of all instructions [Gross et al. 1988]. It is not surprising to discover that, when the processor attempts to execute more than one instruction per cycle, it experiences the greatest number of conflicts over the ALU. The final observation from Figure 3-6 is that out-of-order issue has the most benefit when size of the instruction window is in the range of 16 to 32 instructions.

There is some benefit to duplicating the load/store unit, but this benefit is hard to justify. The data cache is one of the most expensive resources in the processor.

Duplicating this resource requires a large, two-port memory and doubles the logic for detecting a cache hit (tags, comparators, etc.). Also, we have measured the benefit of duplication under optimistic conditions regarding instruction parallelism–the actual benefit is lower. In line with our goal of having a "reasonable" standard processor, we will assume that there is only one load/store unit. To repeat a point made in Section 2.1.3, the presence of resource conflicts in the load/store unit indicates that the data cache is being more fully utilized than in the scalar processor. Increasing the machine parallelism, by duplicating the ALU with out-of-order issue and register renaming, increases the utilization of expensive processor resources–the data cache and memory interface.

3.4   THE DESIGN OF THE STANDARD PROCESSOR

We are now in a position to specify more completely the standard processor we will use to make performance comparisons. The superscalar model presented in this section was culled from a variety of hardware proposals, and its characteristics are based on the observations of the previous section. This model is a good starting point, because it is rather ambitious and can exploit much instruction parallelism. Since it does not limit performance appreciably, it allows a more straightforward exploration of performance limits and design trade-offs than would a processor having limited performance potential.

This section describes the rationale behind the standard processor model as well as the processor itself. There are many possible superscalar organizations, but we are interested in an organization that can be compared to a well-characterized scalar processor. The superscalar processor should mirror the scalar processor as much as possible. Because the processor presented here is intended to aid our measurements, it need not be simple. We are not trying to establish a universal standard or a design that we claim is "best," but rather are trying to establish a basis from which we can explore, refine, and simplify the superscalar processor.

3.4.1   Basic Organization

Figure 3-7 shows a block diagram of the standard superscalar processor. The processor incorporates two major operational units, an integer unit and a floating-point unit, which communicate data via the data cache and data memory. This organization mimics the organization of the R2000 processor and the R2010 floating-point unit, except that there are two integer ALUs in the superscalar processor instead of one. The integer and floating-point units are controlled by a single instruction stream supplied by the instruction memory via the instruction cache. The instruction cache can supply multiple instructions per fetch–the actual number of instructions fetched can be varied.

In Figure 3-7, connections between components should not be construed as single buses; almost all connections comprise multiple buses. Some components of this

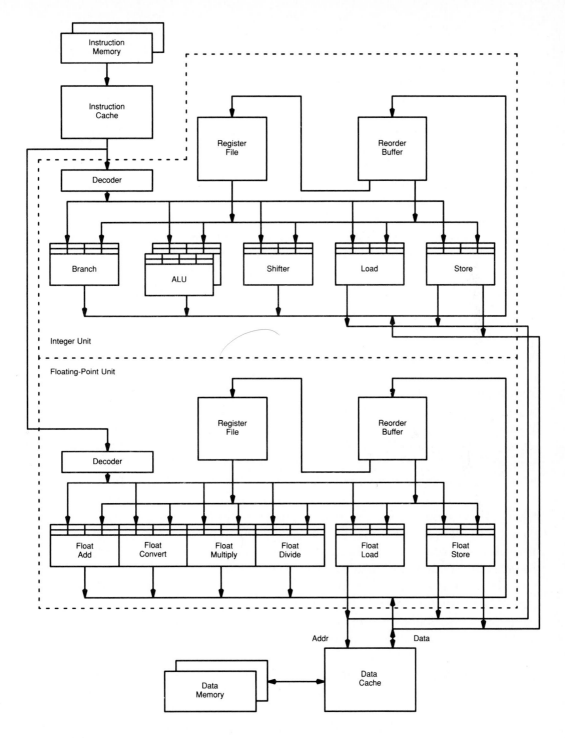

Figure 3-7. Block Diagram of Processor Model

processor are familiar. Other components–such as the storage at the input of the functional units and the *reorder buffer*–are not. The next few sections describe the noteworthy features of this processor.

3.4.2 Out-of-Order Issue

As we saw in Section 3.3.3, out-of-order issue yields the best superscalar perform-ance. Also, as we saw in Section 2.2.3, out-of-order issue requires a buffer, called an instruction window, between the instruction decoder and the functional units. The decoder places instructions into the window, and the instruction-issue logic exam-ines instructions in the window to select appropriate instructions for issue to the functional units.

There are two ways to implement the instruction window. The first is to central-ize the window, buffering every instruction for every functional unit in a common window. The second is to distribute individual buffers to each of the functional units, buffering instructions destined for a particular functional unit at the input of that functional unit. In the latter approach, the buffers are called *reservation stations* [Tomasulo 1967]. As shown in Figure 3-7, the model we will use has a reservation station at the input of every functional unit. Reservation stations were chosen, rather than a centralized window, for reasons to be given shortly.

Figure 3-8 illustrates that the performance of a superscalar processor results from a large variation in parallel-issue cycles. Figure 3-8 plots–for the sample benchmarks *ccom*, *irsim*, *troff*, and *yacc*–the fraction of total cycles in which 0, 1, 2, 3, and so on instructions are issued, in a processor with perfect instruction and data caches, a 32-entry window, and an additional ALU (this was one of the configura-tions shown in Figure 3-6; the number of zero-issue and single-issue cycles cannot be compared to those in Figure 3-2 because an infinite number of functional units was assumed in Figure 3-2 whereas only the ALU is duplicated in Figure 3-8). Figure 3-8 also shows the net instruction-execution rate (not the speedup, as in Figure 3-6) for these benchmarks. These data illustrate that the maximum number of instructions issued in one cycle can be significantly higher than the average instruc-tion-execution rate–more than a factor of two higher.

To avoid as many constraints as possible, the execution unit should support a high peak instruction-issue rate of five or more instructions per cycle. However, is-suing many instructions in one cycle can be quite expensive, because of the cost in-curred in communicating operand values to the functional units. Each instruction issued in a single cycle must be accompanied by all of its required operands, and each instruction can have as many as two operands, so issuing N instructions in one cycle requires access ports and routing buses for as many as $2N$ operands. With a central window, this busing is routed to all functional units. On the other hand, if the instruction window is distributed as reservation stations, the reservation stations need only be filled at the average instruction-execution rate, rather than the peak

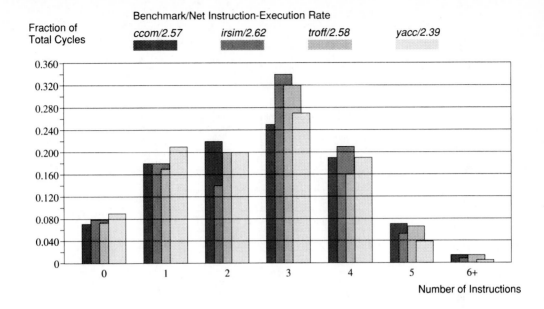

Figure 3-8. Sample Instruction-Issue Distribution

rate; the average rate is between two and three instructions per cycle. Reservation stations potentially reduce the number of operand buses routed to all functional units, because the average instruction-execution rate is much lower than the peak rate.

Because the reservation stations are distributed, they can easily support the maximum instruction-issue rate. In one cycle, the reservation stations can issue an instruction at each of the functional units. The operands needed for instruction issue come from the local reservation stations, rather than being communicated over global (chipwide) buses. At this point in our investigation, we do not know whether this organization really is necessary, because we do not know the penalty of not supporting the peak rate. We simply want to avoid this question this early on. We will examine out-of-order issue in detail in Chapter 7.

The reservation stations of the functional units do not all have to be of the same size. Table 3-3 shows the sizes selected for our model. These sizes were determined by experimentation. Note that the instruction window has a total of 38 entries. This is bigger than the sizes shown in Figure 3-6 because a central instruction window was assumed in Figure 3-6. If the window is partitioned among the functional units, the total size of the window must be larger to support the same amount of lookahead.

To summarize the operation of the processor:

Table 3-3. Sizes of Reservation Stations

Functional Unit	Reservation Station Entries
Integer ALU (2)	4
Barrel Shifter	2
Branch Unit	4
Load Unit	8
Store Unit	8
Float Add	2
Float Multiply	2
Float Divide	2
Float Convert	2

- A multiple-instruction decoder places instructions and operands into the reservation stations of the appropriate functional units.

- A functional unit can issue an instruction in the cycle following decode if that instruction has no dependencies and the functional unit is not busy.

- If the instruction does have dependencies or the functional unit is busy, the instruction is stored in the reservation station until all dependencies are released and the functional unit is available.

- If more than one instruction in the reservation station is ready to be issued in the same cycle, the processor selects the one that appeared first in the instruction sequence.

3.4.3 Register Renaming

As we saw in Section 2.2.4, register renaming eliminates storage conflicts (anti- and output dependencies) for registers. To implement register renaming, each operational unit incorporates a *reorder buffer* [Smith and Pleszkun 1985] containing a number of storage locations that are dynamically allocated to instruction results. When an instruction is decoded, its result value is assigned a reorder-buffer location, and its destination-register number is associated with this location; this renames the destination register to the reorder-buffer location. A *tag*, or temporary hardware identifier, is created by hardware to identify the result, and the tag is also stored in the assigned reorder-buffer location. When a subsequent instruction refers to the renamed destination register, in order to obtain the value considered to be stored in the

Superscalar Microprocessor Design

register, the instruction obtains instead the value stored in the reorder buffer or the tag for this value if the value has not yet been computed.

The reorder buffer is a *content-addressable* memory. That is, an entry in the reorder buffer is identified by specifying something that the entry *contains*, rather than by identifying the entry directly. In this case, the entry is identified using the register number that has been written into it. When a register number is presented to the reorder buffer, the reorder buffer provides the latest value written into the register (or a tag for the value if the value is not yet computed). This organization mimics the register file, which also provides a value in a register when it is presented a register number. However, the reorder buffer and the register file use very different mechanisms for accessing the value. The reorder buffer compares the register number to the register numbers in all entries, and returns the value (or tag) in the entry that has a matching register number (we will see shortly what happens when no entry matches)–this is called *associative lookup*. The register file simply decodes the register number and provides the value in the selected entry.

When an instruction is decoded, the register numbers of its source operands are used to access the reorder buffer and the register file at the same time. If the reorder buffer does not have an entry whose register number matches the source-register number, then the value in the register file is selected as the operand. If the reorder buffer does have a matching entry, the value in this entry is selected because this value must be the value most recently assigned to the register. If the value is not available, because it has not been computed yet, the tag for the value is selected. In any case, the value or tag is copied to the reservation station. This procedure is carried out for each operand required by each decoded instruction.

In a normal instruction sequence, a given register may be written many times. Thus it is possible that different instructions cause the same register number to be written into different entries of the reorder buffer, because the instructions specify the same destination register. To obtain the correct value in this case, the reorder buffer prioritizes multiple matching entries by order of allocation, and returns the most recent entry. In this manner, new entries supersede older entries. An alternative solution is simply to discard each old, superseded value and free the corresponding reorder-buffer entry so that it does not participate in the associative lookup. However, the old entry is preserved in this model because the entry allows the processor to recover the state associated with in-order completion, enabling the processor to implement precise interrupts.

When a result is produced, it is written to the reorder buffer and to any reservation-station entry containing a tag for this result (this requires comparators in the reservation stations). When the value is written into the reservation stations, it may free one or more waiting instructions to be issued by providing a needed input operand. After the value is written into the reorder buffer, subsequent instructions continue to fetch the value from the reorder buffer–unless the entry is superseded by a new value–until the value is retired by writing it to the register file. Retiring occurs in the order of the original instruction sequence, preserving the in-order state for interrupts and exceptions.

The integer reorder buffer has 16 entries, and the floating-point reorder buffer has 8 entries (see Figure 3-7). Both the integer and the floating-point reorder buffers can accept two computed results per cycle and can retire two results per cycle to the corresponding register file. The integer and floating-point reorder buffers and register files have a sufficient number of read ports to satisfy all instructions decoded in one cycle. As with the sizes of the reservation stations, these characteristics were chosen by experimentation.

At this point, we might ask: Why implement renaming this way? Renaming can be implemented in a more literal sense, using a hardware *mapping table* [Keller 1975]. The mapping table is simply an array accessed by register number, much like the register file. However, instead of storing data values, the mapping table stores identifiers for other registers. When a register number is presented to the mapping table, the mapping table produces the identifier for the hardware register that currently has this number.

The reason we are not using a mapping table in our model is to keep the design of the superscalar pipeline very close to the design of the scalar pipeline. This keeps the execution rate of serial instructions in the superscalar processor as high as the execution rate in the scalar processor. Table 3-4 shows the correspondence between the pipelines of the scalar processor and the superscalar processor. In essence, the reorder buffer augments the register file and operates concurrently with it. The reservation stations replace the pipeline latches at the inputs of the functional units. The distribution of operands and writing of results are similar for both processors, except that the superscalar processor requires more hardware, such as buses and register-file ports.

A register-mapping table would introduce another pipeline stage, between the decode stage and the execute stage, to access the mapping table. The reorder buffer avoids this additional stage, and thus avoids introducing effects that cloud the comparison of the scalar and superscalar processors. Chapter 6 examines other alternatives to renaming.

3.4.4 Loads and Stores

Most RISC instructions read and modify registers, but loads and stores read and modify external memory locations also. We can have true, anti-, and output dependencies on memory locations just as we do on registers. However, memory addresses are quite large–32 bits in a typical RISC processor–and this makes it very difficult to implement renaming. Furthermore, renaming external memory locations does not help performance very much. The number of dependencies and storage conflicts on external memory locations is not as large as on the registers, because loads and stores

---

Table 3-4. Comparisons of Scalar and Superscalar Pipelines

Pipeline Stage	Scalar Processor	Superscalar Processor
Fetch	fetch one instruction	fetch multiple instructions
Decode	decode instruction	decode instructions
	access operands from register file	access operands from register file and reorder buffer
	copy operands to functional-unit input latches	copy operands to functional-unit reservation stations
Execute	execute instruction	execute instructions
		arbitrate for result buses
Writeback	write result to register file	write results to reorder buffer
	forward results to functional-unit input latches	forward results to functional-unit reservation stations
Result Commit	n/a	write results to register file

comprise only about 25-30% of the instructions [Gross et al. 1988] and read or write a single memory location, whereas the majority of instructions read two registers and write one register. Moreover, because memory locations are much more abundant than registers, memory locations are not reused in the sense that registers are, and reuse is not as frequent.

In our model, even though loads and stores can be decoded at the same time and placed into their respective reservation stations, only one load or store is issued per cycle from the reservation stations to the data cache, over a single interface (there are separate buses for addresses and data). A load is given priority over a store to use the data-cache interface, because the load is likely to produce a value that the processor needs to proceed with computation. If a store conflicts with a load for the data-cache interface, the store is held in a *store buffer* until the store can be performed. Furthermore, a store is performed in program-sequential order with respect to other stores, and is performed only after all previous instructions, including loads, have completed. This preserves the processor's in-order state in the data cache, because cache updates are performed in the order specified by the program, and because no cache update is performed until it is absolutely correct to do so. The store buffer aids in keeping stores in the correct order and in deferring the completion of a store until previous instructions have completed. In the simulation model, the store buffer has eight entries.

Because stores are held until the completion of all previous instructions, and because loads produce values needed for computation in the processor, keeping loads in program order with respect to stores has a significant negative impact on performance. If a load waits until all preceding stores complete, and therefore waits until all instructions preceding the most recent store complete, then all instructions following the load that depend on the load data also wait. To avoid this performance problem, a load is allowed to bypass preceding stores that are waiting in the store buffer, and the load data is allowed to be used in subsequent computation.

When a load can bypass previous stores, the load may need to obtain data from a previous store that has not yet been performed. The processor checks for a true dependency that a load may have on a previous store by comparing the *virtual* memory address of the load against the virtual memory addresses of all previous, uncompleted stores (the virtual addresses are the addresses computed directly by the instructions, before address translation by a memory-management unit has been applied). For this model, it is assumed that there is a unique mapping for each virtual address, so that it is not possible for two different virtual addresses to access the same physical location. With this assumption, virtual-address comparisons detect all dependencies between physical memory locations. A load has a true dependency on a store if the load address matches the address of a previous store, or if the address of any previous store is not yet computed (in this case, the dependency cannot be detected, so the dependency is assumed to exist). If a load is dependent on a store, the load cannot be satisfied by the data cache, because the data cache does not have the correct value. If the valid address of a store matches the address of a subsequent load, the load is satisfied directly from the store buffer–once the store data is valid–rather than waiting for the store to complete.

As the foregoing discussion implies, loads and stores are performed at the data cache in a manner that avoids anti- and output dependencies on memory locations. Loads can bypass previous stores, but a store cannot bypass previous loads, so there can be no antidependencies between loads and stores. A store is issued in program order with respect to other stores, so there can be no output dependencies between stores.

Loads are performed at the data cache in program order with respect to other loads. While this is unnecessary for correctness, it does simplify the implementation of the load reservation station, as we will see. In addition, there is little or no performance advantage to allowing loads to be performed out of order at the data cache, because the data supplied to the processor by older loads is more likely to be needed in computation than the data supplied by newer loads.

This model also assumes that loads do not change the permanent state of the data cache or main memory, so that loads cannot give rise to dependencies within the cache or memory. In general, this restriction is easily met in a cache-based system. Loading cacheable data, by definition, cannot change the state of the system. Furthermore, the model assumes that an unnecessary or incorrect load can be performed without harming the system–for example, that a load of cached data can be performed even though a previous instruction has an (as yet undiscovered) exception. This restriction is also easily met in a cache-based system, because a cache reload may perform unnecessary loads and may repeatedly load the same data.

3.4.5 The Performance of the Model

To finish the presentation of the superscalar processor model, Figure 3-9 shows its performance. Figure 3-9 shows the speedup of the model using the three instruction-issue policies described in Section 2.2, to illustrate the benefit of out-of-order issue (and out-of-order completion) under more realistic assumptions than in Figure 3-6. The presence of reservation stations in the model does not imply that the model must perform out-of-order issue. This capability can be disabled.

With in-order issue, out-of-order completion yields higher performance than in-order completion. In fact, with in-order completion, the superscalar processor has lower performance than the scalar processor for some floating-point-intensive programs; using pipelined floating-point units and out-of-order completion, the scalar

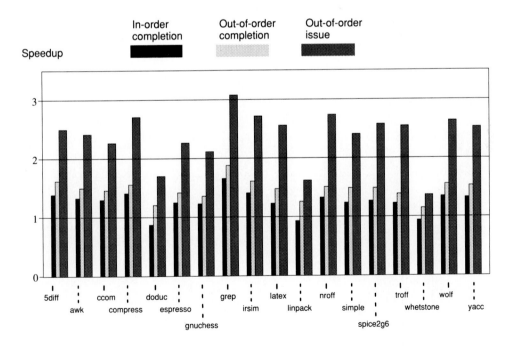

Figure 3-9. Potential Speedup of Three Instruction-Issue Policies,
Using Ideal Instruction Fetcher

processor can take some advantage of instruction parallelism. Still, out-of-order issue consistently has the best performance for the benchmark programs.

However, we have been conveniently ignoring procedural dependencies for some time now. The performance in Figure 3-9 is measured using ideal instruction fetching. The instruction trace can supply as many instructions in one cycle as the processor can consume, except when an instruction-cache miss occurs, with little regard for branches. Of course, this ability is not available to any real processor–it is available to us only because the simulator analyses a trace *after* execution. We are at a point where procedural dependencies must be considered.

3.5 THE REAL PERFORMANCE LIMIT: PROCEDURAL DEPENDENCIES

A scalar processor, upon decoding a branch instruction, waits until the execution of the branch before fetching the next sequence of instructions. Figure 3-10 shows what happens to the performance of the superscalar processor if it uses the same approach (Figure 3-10 uses the convention–followed throughout the remainder of this book–of showing the low, harmonic mean, and high speedups of all benchmarks). Under these conditions, out-of-order issue has much less advantage than is implied by Figure 3-9 (Figure 3-10 also summarizes the performance of Figure 3-9 for com-

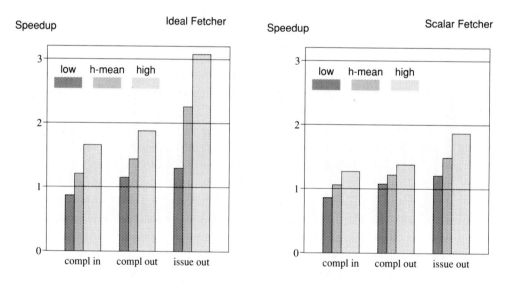

Figure 3-10. Speedups with an Ideal Instruction Fetcher and with the Instruction Fetcher Modeled After a Scalar Fetcher

parison). The average speedup with out-of-order issue and an ideal fetcher is 2.4; the speedup decreases to 1.5 with a scalar fetcher.

For general-purpose applications, the penalties incurred for procedural dependencies by a conventional instruction fetcher reduce the processor's ability to have sufficient instructions to execute in parallel. Unless these penalties can be reduced, out-of-order issue is not worth the effort. The next chapter addresses this problem, before we proceed with refining the execution hardware. Unless instruction fetching can be made more efficient, there is not much point to further refinements.

3.6   BACKGROUND

Many of the ideas presented in Chapters 4 through 9 are from the published literature on out-of-order issue, register renaming, precise interrupts, and multiple-instruction issue. This section briefly reviews this literature for readers interested in the historical progression of ideas related to superscalar hardware.

Tomasulo's paper [1967] on instruction issuing in the IBM 360/Model 91 is the classical reference for out-of-order issue with register renaming. Weiss and Smith [1984] examine the use of Tomasulo's algorithm on the CRAY-1 scalar unit. Patt et al. [1985a] and Hwu and Patt [1986] incorporate Tomasulo's algorithm into a proposal that, in concept, issues multiple microinstructions out of order. Pleszkun and Sohi [1988] examine multiple-instruction issue in the CRAY-1 scalar unit. Murakami et al. [1989] examine Tomasulo's algorithm on a RISC-like processor.

Tjaden and Flynn [1970] published the first paper, to my knowledge, on multiple-instruction issue. This work is carried on (in a rather theoretical vein) by Tjaden [1972], Tjaden and Flynn [1973], Wedig [1982], and Uht [1986].

Torng [1984] focuses on a specific hardware proposal for multiple-instruction issue: the *dispatch stack*. Acosta et al. [1986] is a later version of this. Dwyer and Torng [1987] examine the hardware cost of the dispatch stack.

Smith and Pleszkun [1985] provide a very complete survey of techniques that provide precise interrupts. Pleszkun et al. [1987] extend these ideas in a software-oriented approach. Sohi and Vajapeyam [1987] relate the techniques of Smith and Pleszkun [1985] to Tomasulo's algorithm and register renaming.

Agerwala and Cocke [1987] made the earliest reference to term *superscalar* that I can find. This paper is largely about pipelined RISC processors, but in the final sections it contains interesting comparisons of superscalar and vector architectures. This paper apparently spawned the IBM RIOS architecture and the IBM RISC System/6000™ line of workstations [Groves and Oehler 1989].

More recent studies of superscalar processors examine the limitations on superscalar performance [Jouppi 1989a, Smith et al. 1989, Johnson 1989] and the benefits that software can provide [Wulf 1988, Jouppi and Wall 1988, Smith et al. 1990].

Chapter 4

Instruction Fetching and Decoding

Out-of-order issue is effective only when instructions can be supplied at a sufficient rate to keep the execution unit busy. If the average rate of instruction fetching is less than the average rate of instruction execution, performance is limited by instruction fetching. It is easy to provide the required instruction bandwidth for sequential instructions, because the fetcher can simply fetch several instructions per cycle, in blocks of multiple instructions. It is much more difficult to provide instruction bandwidth in the presence of nonsequential fetches caused by branches.

This chapter describes the problems related to instruction fetching in the superscalar processor and proposes solutions to these problems. The technique of branch prediction allows an adequate instruction-fetch rate in the presence of branches and is key to achieving performance with out-of-order issue. Furthermore, even with branch prediction, performance is highest with a wide instruction decoder (four instructions wide). Only a wide decoder can provide sufficient instruction bandwidth for the execution unit, because of the misalignment of branch-target instructions in memory.

4.1 BRANCHES AND INSTRUCTION-FETCH INEFFICIENCIES

Branches impede the ability of the processor to fetch instructions because they make instruction fetching dependent on the results of instruction execution. When the outcome of a branch is not known, the instruction fetcher is stalled or may be fetching incorrect instructions, depleting the instruction window of instructions and reducing the chances that the processor can find instructions to execute in parallel. This effect is similar to the effect of a branch on a scalar processor, except that the penalty is greater in a superscalar processor because the superscalar processor is attempting to fetch and execute more than one instruction per cycle.

But branches also affect the execution rate in another way that is unique to the superscalar processor. Branches disrupt the sequentiality of instruction addressing, causing instructions to be misaligned with respect to the instruction decoder. This misalignment in turn causes some otherwise valid fetch and decode cycles to be only partially effective in supplying the processor with instructions, because the entire width of the decoder is not occupied by valid instructions.

The sequentially fetched instructions between branches are collectively called an *instruction run*, and the number of instructions fetched in a run is called the *run length*. Figure 4-1 shows two instruction runs occupying four instruction-cache blocks (recall that we are assuming four-word cache blocks). The first run consists of instructions *S1-S5*, which include a branch to the second run consisting of instructions *T1-T4*. Figure 4-2 shows how these instruction runs are sequenced through straightforward four-instruction and two-instruction decoders assuming–for illustration–that two cycles are required to determine the outcome of a branch.

The *branch delay* of a processor is the delay from the time that the processor decodes a branch instruction to the time that it decodes the first target instruction. The branch delay is one cycle in a typical RISC processor (such as the R2000). The two-cycle branch delay shown in Figure 4-2 is more representative of the branch delay in a superscalar processor. The longer branch delay of the superscalar processor, again relative to the cycle in which the branch is decoded, is a direct result of the wider instruction fetch and decode path. For example, consider in Figure 4-2 that instruction *S4* compares two values and that the outcome of branch *S5* is based on this comparison. Even though *S4* and *S5* are fetched in a single cycle, the processor requires two cycles to execute this sequence: one cycle for *S4* and one cycle for *S5*.

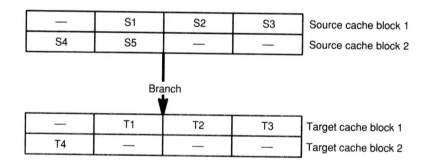

Figure 4-1. Sequence of Two Instruction Runs for Illustrating Fetch Behavior

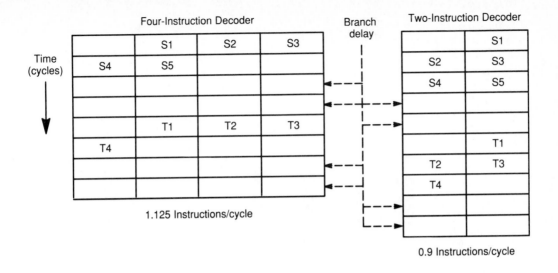

Figure 4-2. Sequence of Instructions in Figure 4-1 Through
Two-Instruction and Four-Instruction Decoders

The branch delay is longer in the superscalar processor because the superscalar processor fetches the branch instruction earlier and because the branch outcome depends on the execution of other instructions.

In the example of Figure 4-2, the four-instruction decoder provides higher instruction bandwidth–1.125 instructions per cycle, compared to 0.9 instructions per cycle for the two-instruction decoder–but neither decoder provides adequate bandwidth to exploit the available instruction parallelism. There are two reasons for this. First, the decoder is idle while the processor determines the outcome of each branch in the sequence. Second, instruction misalignment prevents the decoder from operating at full capacity even when the decoder is processing valid instructions. For example, a four-instruction decoder spends two cycles processing the five instructions *S1-S5*, even though the capacity of the decoder is eight instructions in these two cycles.

The instruction fetcher operates most efficiently when it is processing long runs of instructions. Unfortunately, general-purpose programs, such as those we are using, have instruction runs that are generally quite short. Figure 4-3 shows the distribution of run length for the four sample benchmarks. In Figure 4-3, the run length is determined only by taken branches. Branches that are not taken are counted as part of a longer run. Also, Figure 4-3 shows a dynamic measurement. If a branch is taken more than once during execution, the run length is determined each time and counted as part of the distribution (the run length may or may not be the same each time the

---

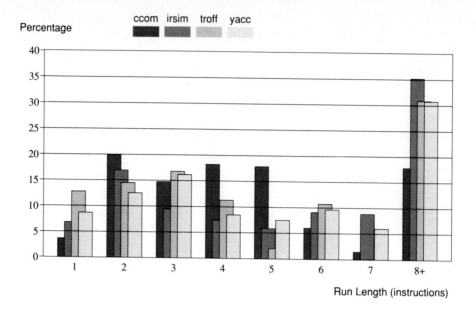

Percentage

ccom irsim troff yacc

Run Length (instructions)

Figure 4-3. Dynamic Run-Length Distribution of Taken Branches

branch is executed depending on the outcomes of preceding branches). The R2000 architecture requires that *no-op* instructions be used to implement pipeline stalls, but these are not counted in the run length.

Taken branches are typically less than 20% of the dynamic instructions executed by a RISC processor [Gross et al. 1988]. The *average* run length in Figure 4-3, about six instructions, is consistent with this measurement, because the average run length takes into account the actual length of each run. There are many runs of eight instructions or more that bring up the average run length. Unfortunately, we are concerned more with the relative mix of long and short runs, rather than the average run length. In the distribution of Figure 4-3, *half* of the instruction runs consist of four instructions or less. The instruction runs shown for the purpose of illustration in Figure 4-2 are representative of the runs in the benchmark programs. The large number of short runs places many demands on the instruction fetcher, if adequate fetch bandwidth is to be achieved.

4.2 IMPROVING FETCH EFFICIENCY

This section discusses mechanisms which improve the instruction-fetch efficiency of the superscalar processor by reducing the penalty of branch delays and instruction misalignment. We also will determine the benefit to processor performance of these mechanisms. Subsequent sections consider hardware implementations.

4.2.1 Scheduling Delayed Branches

A common technique for dealing with the branch delay in a scalar RISC processor is to execute branch instructions prior to the actual change of program flow, using *delayed branches* [Gross and Hennessy 1982]. When a delayed branch is executed, the processor continues execution in the current stream for a number of instructions, while it initiates the fetch of the target instruction-stream (see Figure 4-4). Using this technique, the processor can perform useful computation during the branch delay. The desirable number of instructions executed after the branch, but before the target, is implementation dependent (but this number usually gets ingrained as part of the architecture). The number of delay instructions is chosen according to the anticipated number of cycles required to determine the outcome of a branch and fetch the target instruction. Figure 4-4 shows two branch-delay instructions; in a RISC processor, there is typically a single branch-delay instruction because the branch outcome is determined during the decode stage.

A compiler or other code-generation software takes advantage of delayed branches by *scheduling* instructions after the branches: that is, by attempting to place useful instructions after the delayed branches so that these instructions are executed while the processor determines the branch outcome. The instruction(s) following a given branch is (are) taken either from before the branch, from the target of the branch, or from instructions sequentially following the branch. In the most straight-

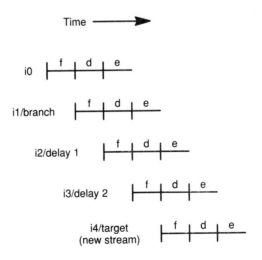

Figure 4-4. Execution of a Delayed Branch

---

forward approach, software cannot schedule instructions in the branch delay unless it is completely safe to do so. However, this approach is effective only if the branch delay is a single cycle, because only one safe instruction can be found most of the time [McFarling and Hennessy 1986].

In a superscalar processor, software scheduling of delayed branches is not very effective. The difficulty presented by the superscalar processor is that, as we have already seen, the branch delay of a superscalar processor is larger than the delay in a scalar processor. The branch delay of a superscalar processor is determined mostly by the time required to determine the outcome of a branch: the more the processor is looking ahead, the longer this time is. In contrast, the branch delay of a scalar processor is determined mostly by the time taken to fetch the target instruction: this can be made as small as one cycle with straightforward techniques. In a superscalar processor with out-of-order issue, the branch delay can be quite long. Also, because the superscalar processor fetches more than one instruction per cycle, a very large number of instructions must be scheduled after the branch to overcome the penalty of the branch delay.

We argued in Section 4.1 that the longer branch delay of a superscalar processor is a direct result of its lookahead ability. Figure 4-5 presents actual measurements of this effect. Figure 4-5 plots extremes, over the sample benchmarks, of the average magnitude of the branch delay as a function of speedup. Figure 4-5 also shows the

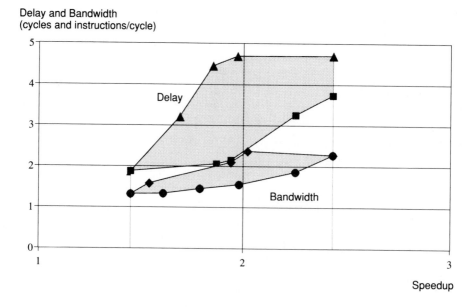

Figure 4-5. Branch Delay and Penalty Versus Speedup

extremes of instruction-fetch bandwidth sustained during the time interval of the branch delay. The range of speedups was obtained by varying machine configurations and instruction-fetch mechanisms over all the sample benchmarks.

Figure 4-5 illustrates that the branch delay increases significantly as the processor lookahead, and thus speedup, increases. The increase of the branch delay with increasing speedup is the result of the buffering provided by the instruction window. With larger speedups, the instruction window is kept relatively fuller than with smaller speedups. A full window, and the dependencies a full window implies, prevent the resolution of branches for several cycles–during these cycles, the processor is looking ahead beyond the branch to find instructions to execute. Furthermore, during these cycles, the processor must sustain an instruction-fetch rate of about two instructions per cycle to support the lookahead. Thus, for example, to sustain a speedup of two, software would have to schedule about six to eight instructions following a branch (three to four cycles of delay at two instructions per cycle). It is very unlikely that software would be able to schedule this many instructions. Furthermore, the instructions scheduled in the branch delay would be likely also to contain one or more branches, because instruction runs are typically short. It is difficult to define what the hardware should do when branches appear as delay instructions of another branch, especially because these other branches may also have delay instructions. There are too many cases to consider to have much hope of designing the correct behavior in every case [Pleszkun et al. 1987].

Software techniques for overcoming branch delays in a superscalar processor must be more sophisticated than the techniques used to schedule delayed branches in a scalar RISC processor. In Chapter 12, we will consider a more powerful technique that relies on more hardware support than simply delaying the branch.

4.2.2 Branch Prediction

Hardware can reduce the average branch delay by predicting the outcomes of branches during instruction fetching, without waiting for the execution unit to indicate whether or not the branches should be taken. Figure 4-6 illustrates how this helps for the instruction sequence of Figure 4-1. Branch prediction relies on the fact that the future outcome of a branch usually can be predicted from previous branch outcomes. *Hardware branch-prediction* predicts branches dynamically using a structure such as a *branch-target buffer* [Lee and Smith 1984] and relies on the fact that the outcome of a branch is stable over periods of time. *Software branch-prediction* predicts branches statically by annotating the program with prediction information (such as by setting a bit in a branch instruction to indicate the outcome of the branch) and relies on the fact that the outcome of a branch usually favors one possible outcome over the other. The favored outcome is determined by measuring an instruction trace or by making a reasonable guess based on program context (for example, loop branches are likely to be taken more often than not).

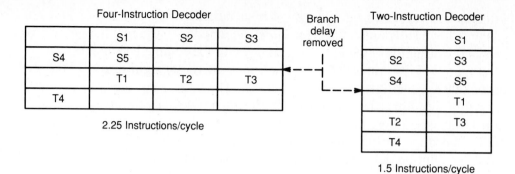

Figure 4-6. Sequence of Instructions in Figure 4-1 Through Two-Instruction and Four-Instruction Decoders with Branch Prediction

It is important that branch prediction be accurate, because the processor suffers the full branch-delay penalty if the prediction is wrong. Though they rely on different branch properties, software and hardware branch-prediction yield comparable prediction accuracy [McFarling and Hennessy 1986, Ditzel and McLellan 1987]. It is further important, however, that the branch-prediction algorithm not incur any delay cycles to determine the predicted branch outcome. With reference to Figure 4-6, if an additional cycle is required between runs to determine the predicted outcome of the branch, the fetch efficiency of the four-instruction decoder drops to 1.5 instructions per cycle, and the fetch efficiency of the two-instruction decoder drops to 1.125 instructions per cycle. An additional cycle between instruction runs causes a large reduction in fetch efficiency, because instruction runs are generally short. Hardware branch-prediction has some advantage over software prediction in that it easily avoids this additional cycle (though at the expense of additional hardware complexity). Software prediction requires an additional cycle for decoding the prediction information and computing the target address, even if the instruction format explicitly describes the prediction and the absolute target addresses (for example, using a single instruction bit for the prediction and a fixed field containing the absolute target address). Decoding the branch instruction, extracting the target address from the instruction, and applying this address to the cache still take more time than if hardware discovers the branch and predicts its outcome during the fetch stage.

Hardware that uses branch prediction overcomes the penalty of branch delays more effectively than software scheduling of delayed branches. The branch-prediction information can be supplied either by hardware or software, but the instruction fetcher should be designed to fetch instruction runs without any intervening delay cycles. Furthermore, to support lookahead, the fetcher should be prepared to fetch and decode several runs before determining the outcome of a given branch. For ex-

Superscalar Microprocessor Design

ample, considering that the median run is four instructions long, a four-instruction fetcher might fetch up to four complete instruction runs to obtain a speedup of two, because the branch delay can be more than four cycles (note that this implies a window of about 16 entries to buffer the instruction runs).

Because the scope of this book does not permit the extensive evaluation of software techniques, hardware branch-prediction will be used in our evaluations. Section 4.3 describes the branch-prediction algorithm we will use.

4.2.3 Aligning and Merging

In addition to improving the instruction-fetch efficiency by predicting branches, the instruction fetcher can improve efficiency by aligning instructions for the decoder, so that there are fewer wasted decoder slots. In the absence of branch-delay penalties, the ratio of wasted to used decoder slots depends on the particular alignment of an instruction run in memory (and, therefore, the alignment in the instruction cache). Figure 4-7 demonstrates the effect of instruction alignment on fetch efficiency, comparing the fetch efficiency of a two-instruction decoder with that of a four-instruction decoder. These fetch efficiencies were determined during execution of the sample benchmarks, excluding all branch-delay cycles; only misalignment causes the fetch efficiencies to be fewer than two or four instructions per cycle.

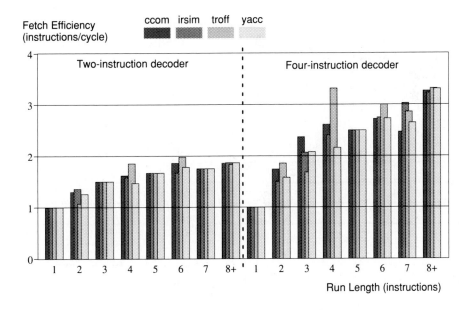

Figure 4-7   Fetch Efficiencies for Various Run Lengths

Both decoders approach their maximum throughputs of two and four instructions per cycle, respectively, as the runs become longer and longer, but the four-instruction decoder exceeds two instructions per cycle for many of the short runs, whereas the two-instruction decoder does not. For these programs, the average fetch efficiency is 1.72 instructions per cycle for a two-instruction decoder and 2.75 instructions per cycle for a four-instruction decoder. Short instruction runs penalize the two-instruction decoder more than the four-instruction decoder, because misalignment frequently adds one cycle to the fetch time of the two-instruction decoder, whereas the four-instruction decoder can handle a relatively large number of misaligned runs with no additional penalty.

If the fetcher can fetch instructions faster than they are executed (for example, by fetching an entire four-word cache block in a single cycle to supply a two-instruction decoder), it can align fetched instructions to avoid wasted decoder slots. Figure 4-8 shows how the instruction runs of Figure 4-1 are aligned in a two-instruction and a four-instruction decoder (successful branch prediction is assumed for this illustration). Note that aligning has reduced the wasted decoder slots for run *T1-T4*, but has not reduced the wasted decoder slots for run *S1-S5*. Also, note that aligning makes it more difficult to determine the address of an instruction in a given decoder position, thus making it more difficult to implement program-counter-relative branches and to report the addresses of instructions causing exceptions.

If the fetcher performs branch prediction, it can also increase the fetch efficiency by merging instructions from different runs. This is illustrated by Figure 4-9 (the decode slots labeled "next" in Figure 4-9 refer to instructions from the run following *T1-T4*). As with aligning, merging depends on the ability of the fetcher to fetch instructions at a rate greater than their execution rate, so that additional fetch cycles are available when the fetcher reaches the end of an instruction run.

Four-Instruction Decoder

S1	S2	S3	S4
S5			
T1	T2	T3	T4

3 Instructions/cycle

Two-Instruction Decoder

S1	S2
S3	S4
S5	
T1	T2
T3	T4

1.8 Instructions/cycle

Figure 4-8. Sequence of Instructions in Figure 4-1 Through Two-Instruction and Four-Instruction Decoders with Branch Prediction and Aligning

Four-Instruction Decoder

S1	S2	S3	S4
S5	T1	T2	T3
T4	next	next	next

4 Instructions/cycle

Two-Instruction Decoder

S1	S2
S3	S4
S5	T1
T2	T3
T4	next

2 Instructions/cycle

Figure 4-9. Sequence of Instructions in Figure 4-1 Through Two-Instruction and Four-Instruction Decoders with Branch Prediction, Aligning, and Merging

4.2.4 Simulation Results and Observations

The performance of the various fetch alternatives described in Sections 4.2.2 and 4.2.3 is shown in Figure 4-10 and Figure 4-11. The figures plot, for the alternatives, the low, harmonic mean, and high speedups among all benchmark programs. The performance of both a two-instruction and a four-instruction decoder is shown to emphasize the relative misalignment penalties. The interpretation of the chart labels is as follows:

- base – no prediction and no aligning.
- pred – predicting branches with hardware.
- align – aligning the beginning of instruction runs in the decoder when possible.
- merge – merging instructions from different runs when possible.
- perfect – perfect branch prediction, but with alignment penalties. This case is included for comparison.
- max – the maximum speedup allowed by the execution hardware, with no fetch limitations. This case is included for comparison.

The speedups for all but the two final sets of results are cumulative: the speedups for a given case include the benefit of all previous cases. Aligning and merging are performed only when the fetcher has sufficient additional bandwidth, in comparison to the execution rate, to perform these operations. The additional bandwidth is needed because the fetcher must buffer at least two blocks in order to be able to align or merge: in the case of aligning, these blocks are from the same run, and in the case of merging, these blocks can be from different runs.

Speedup

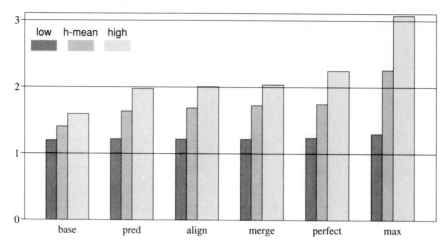

Figure 4-10. Speedups of Fetch Alternatives with Two-Instruction Decoder

Speedup

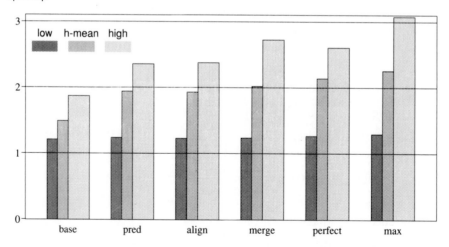

Figure 4-11. Speedups of Fetch Alternatives with Four-Instruction Decoder

The principal observation from these performance measurements is that branch prediction yields the greatest incremental benefit of any of the mechanisms for improving fetch efficiency. Section 4.3 discusses a hardware implementation of branch prediction that incurs a relatively small hardware cost.

Figure 4-10 and Figure 4-11 also show that a four-instruction decoder always outperforms a two-instruction decoder. This is not surprising, because the four-instruction decoder has twice the potential instruction bandwidth of the two-instruction decoder. None of the techniques applied to a two-instruction decoder to overcome this limitation—such as aligning or merging—yields the same performance as a four-instruction decoder. The essential problem with a two-instruction decoder is that the instruction throughput can never exceed two instructions per cycle—the fetch bandwidth is always below this limit. However, a two-instruction decoder places fewer demands on hardware than a four-instruction decoder, particularly for read ports on the register file. Section 4.4 considers the implementation of a four-instruction decoder and the demands it places on hardware.

There is some advantage to aligning and merging instruction runs, but these do not seem to be appropriate functions for hardware, because they increase the time taken by the instruction-fetch logic. The path from the instruction cache to the decoder is a *critical path* in many processors, meaning that the path determines the cycle time. A more efficient method to align and merge instruction runs relies on software. Software can align instruction runs by arranging instructions on appropriate boundaries in memory. Software also can merge instruction runs using software branch-prediction. If software can effectively predict the outcome of a branch, it can move instructions from the likely path to pad the remainder of the decode positions following the branch (these instruction are decoded at the same time as the branch). This is similar in concept to scheduling branch delays in a scalar processor, but the goal is to improve the efficiency of the decoder, and fewer instructions are required to accomplish this—one instruction for a two-instruction decoder and one to three instructions for a four-instruction decoder. The padded instructions are executed if the prediction is correct and are suppressed if the prediction is not correct. The reorder buffer (or similar structure) in the execution unit can easily nullify a variable number of instructions following a branch, depending on the alignment and outcome of the branch: we will look into this more in Chapter 5.

Software aligning and merging has the advantage that instruction runs are always aligned and are merged when a branch is successfully predicted. This is in contrast to hardware, which can perform these functions only with additional fetch bandwidth and with alignment hardware in the critical path between the instruction cache and decoder. However, software merging, since it relies on software prediction and on branches with a variable number of delay instructions, is not possible with any existing instruction set.

4.2.5 Multiple-Path Execution

Our philosophy for branch prediction has been to use branch prediction to determine the likely outcome of a branch and to have the processor pursue the likely execution path. At any point in execution, if the hardware determines that a prediction was

---

incorrect (we will see more about how this is accomplished in the next section), the hardware backs up in the instruction stream (using mechanisms that provide precise interrupts, as we will see in the next chapter) and proceeds down the correct path. The branch penalty manifests itself through the instructions completed after the branch is predicted but before the branch outcome is determined. The results of these instructions are discarded, and the time that the processor spent executing them is wasted.

Theoretically, there is another approach that avoids the branch delay altogether. If we were to have the processor pursue *both* paths at a branch, we could be sure that the processor were executing the correct instruction sequence when it determined the outcome of the branch. The processor would simply discard the results of the incorrect path. In fact, this approach has been used in previous scalar processors to reduce the branch penalty, notably the IBM 370/Model 168 and the IBM 3033 [Lee and Smith 1984]. However, this approach is not one we will explore in detail, because of the hardware costs it incurs.

To pursue efficiently both paths at a branch, the processor must have enough resources to pursue both paths: two instruction-cache ports, double the number of decoders and functional units, and so on. Doubling the processor resources to eliminate the penalty of one branch might be a good trade-off, but the problem is not solved this easily. We must double the resources at *each* branch that is encountered while the processor determines the outcome of the first branch, causing hardware requirements to grow rapidly. Consider typical values we have seen: the branch delay is four cycles, the median run length is four instructions, and the decoder is four instructions wide. The processor might easily encounter three or four more branches while it is waiting to determine the outcome of a branch, meaning that we need 8 or 16 times the resources in order not to suffer a branch penalty due to resource conflicts.

Obviously, providing this amount of hardware to eliminate the branch penalty is difficult to justify. However, consider the alternative: that the processor pursues multiple code paths but with reduced efficiency because the required hardware is not present. If we use prediction to pursue the likely path, this path is executed at 100% efficiency, but it is the incorrect path with a probability depending on the branch-prediction accuracy. If, on the other hand, we pursue both paths at a branch, we always have the correct path in execution, but with lower efficiency depending on the processor resources. In rough terms, let us say that the processor has to sustain execution through three branches, that the branch-prediction accuracy is 85% (a reasonable average), and that, with multiple-path execution, each path from a branch is executed at one-half efficiency since there are two paths. On the average, branch prediction would be correctly executing instructions after the third branch about 61% of the time (.85 cubed), but multiple-path execution would be executing these correct instructions with 12.5% efficiency (.50 cubed). This rough analysis argues strongly that the best alternative is to pursue the most probable path, as determined by branch prediction.

4.3 IMPLEMENTING HARDWARE BRANCH-PREDICTION

A conventional method for hardware branch-prediction uses a branch-target buffer [Lee and Smith 1984] to collect information about the most recently executed branches. Typically, the branch-target buffer is accessed using an instruction address, and the buffer indicates whether or not the instruction at that address is a branch instruction. If the instruction is a branch, the branch-target buffer indicates the predicted outcome and the target address. Figure 4-12 shows the average prediction accuracy of a branch-target buffer on the sample benchmark programs. Figure 4-12 is included only as an illustration: the indicated branch-prediction accuracy agrees with the accuracy reported by others [Lee and Smith 1984, McFarling and Hennessy 1986, Ditzel and McLellan 1987]. A large branch-target buffer is required to achieve good accuracy, because, unlike a cache, there is one entry per address tag and the branch-target buffer cannot take advantage of spatial locality [Smith 1982]. An instruction cache often provides an entire block of instructions for every good cache entry (usually four or more instructions), but the branch-target buffer provides only one prediction. Even with a large number of entries, moreover,

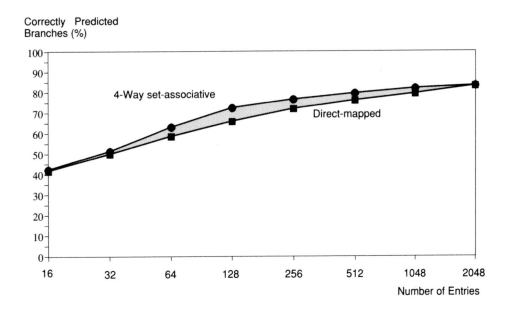

Figure 4-12. Average Branch-Target Buffer Prediction Accuracy

the accuracy of a branch-target buffer is still rather low, because of misprediction. It is not sufficient just to find the branch instruction in the buffer, as with a cache–the prediction also must be correct.

A better hardware alternative to a branch-target buffer is to include branch-prediction information in the instruction cache. Each instruction-cache block can include additional fields that indicate the address of the block's successor (that is, the block to be fetched after the current block) and that indicate the location of a branch in the block, if there is a branch in the block. When the instruction fetcher obtains a cache block containing correct information, it can easily fetch the next cache block without waiting on the decoder or execution unit to indicate the proper fetch action to be taken. This is comparable to a branch-target buffer with a number of entries equal to the number of blocks in the instruction cache and with associativity equal to the associativity of the instruction cache. This scheme is nearly identical to the branch-target buffer in prediction accuracy (most of the benchmark programs experienced a branch-prediction accuracy of 80-95% with a 64-Kbyte cache).

It is common to implement a procedure return using an unconditional, indirect branch: that is, a branch that is always taken and whose target address is specified by the contents of a register. Indirect branches can reduce the branch-prediction accuracy even when they are unconditional, because branch-prediction hardware cannot always predict the target address when this address can change during execution. However, often the target address of an indirect branch does not change over a span of time because of repetitive procedure calls from the same procedure. This makes it useful to predict indirect branches if the cost is not high (the benefit is about a 2-3% improvement in branch-prediction accuracy).

The results given in Section 4.2.4 assumed the existence of an instruction fetcher that can fetch instructions without any delay for successfully predicted branches. This section discusses the implementation of this instruction fetcher to demonstrate that an implementation is feasible and to show that it causes only a small increase in the size of the instruction cache. Also, this section shows that the fetcher can predict indirect branches with no additional cost over that required for the basic implementation.

There are many possible variations on the hardware described here, so we will not delve too deeply into the exact mechanism. Furthermore, even though the simulated performance in Section 4.2.4 correctly accounted for the delayed branches in the R2000 architecture, we will ignore delayed branches in the following discussion, for the same reasons that we are ignoring CISC processors. Delayed branches can be handled using the techniques described shortly, but delayed branches introduce problems that interfere with understanding the basic mechanism. Our objective is to explore fundamental ideas, not to investigate unique problems.

4.3.1 Basic Organization

Figure 4-13 shows a sample organization for the instruction-cache entry required by the instruction fetcher (this entry may include other information not important for this discussion). For this example, the cache entry holds four instructions. The entry also contains instruction-fetch information which is shown expanded in Figure 4-13. The fetch information contains an address tag (used in the normal fashion) and two additional fields used by the instruction fetcher:

- The *successor index* field indicates both the next cache block predicted to be fetched and the first instruction within this next block predicted to be executed. The successor index does not specify a full instruction address, but is of sufficient size to select any instruction within the cache. For example, a 64-Kbyte, direct-mapped cache requires a 14-bit successor index if all instructions are 32 bits in length (12 bits to address the cache block and 2 bits to address the instruction in the block if the block size is four words).

- The *branch block index* field indicates the location of a branch point within the corresponding instruction block. Instructions beyond the branch point are predicted not to be executed.

Figure 4-14 shows sample address tag, successor index, and branch block index entries for the code sequence of Figure 4-1, assuming a 64-Kbyte, direct-mapped cache and the indicated instruction addresses.

4.3.2 Setting and Interpreting Cache Entries

When a cache entry is first loaded, the cache-fetch hardware sets the address tag in the normal manner and sets the successor index field to the next sequential fetch ad-

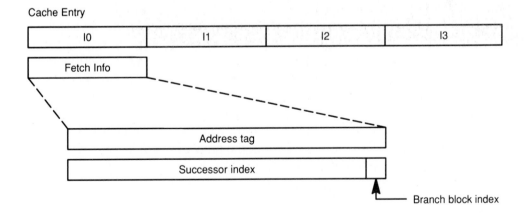

Figure 4-13. Instruction-Cache Entry for Branch Prediction

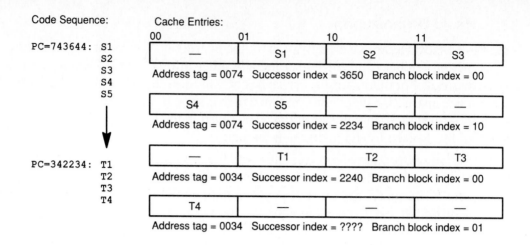

Code Sequence: Cache Entries:

PC=743644: S1
           S2
           S3
           S4
           S5

PC=342234: T1
           T2
           T3
           T4

Figure 4-14. Example Cache Entries for Code Sequence of Figure 4-1

dress. The default for a newly loaded entry, therefore, is to predict that branches are not taken. If the prediction is incorrect, this will be discovered later by the normal procedure for detecting a mispredicted branch.

As Figure 4-14 illustrates, a target program counter can be constructed at branch points by concatenating the successor index field of the branching entry to the address tag of the successor entry. To see this, consider concatenating the successor index of the second cache block (the block that branches) to the address tag of the third block (the target block). This yields the target address of the second run of instructions *T1-T4*. Between branch points, the program counter is incremented and used to detect cache misses for sequential runs of instructions. Cache misses on branch targets are handled as part of branch-prediction checking, as described shortly. The program counter recovered from the cache entries can be used during instruction decode to compute program-counter-relative addresses and procedure return addresses; if the program counter is wrong because of misprediction, this will be detected later. In a set-associative cache, some bits in the successor index field are used to select a block within a cache set, and are not involved in generating the program counter.

The validity of instructions at the beginning of a cache entry is determined by low-order bits of the successor index field in the preceding entry. When the preceding entry predicted a taken branch, this entry's successor index may point to any instruction within the current block, and instructions up to this point in the block are not to be executed. The validity of instructions at the end of the block is determined by the branch block index, which indicates the point where a branch is predicted to be taken. Instructions beyond the branch are not to be executed. The branch block index is used by the instruction decoder to determine the location of valid instructions and is not used by the instruction fetcher to obtain instructions. The instruction fetcher retrieves cache entries based on the successor index fields alone.

4.3.3 Predicting Branches

To check each branch prediction, the processor keeps a list of predicted branches ordered by the sequence in which branches were predicted. This list can be kept in an array, such as a first-in, first-out buffer. Each entry on this list indicates the location in the cache of the corresponding branch; this location is formed from the branching block's program counter and the branch block index, which gives the specific location of the branch within the block. Each entry on the list also contains a complete program-counter value for the target of the branch. Because the cache predicts only taken branches, this list contains only taken branches. Nontaken branches are predicted by the absence of prediction information.

The processor executes all branches in their original program sequence (this is guaranteed by the operation of the branch-execution unit). These branches are detected by instruction decoding, independent of prediction information maintained by the instruction fetcher. When a branch is executed, the processor compares information related to this branch with the information at the front of the list of predicted branches (this is the oldest predicted-taken branch). The following conditions must hold for a successful prediction:

- If the executed branch is taken, its location in the cache must match the location of the next branch on the list of predictions. If a taken branch is detected at an unexpected location during execution, this indicates that the taken branch was predicted to be not taken.

- If the location of the executed branch matches the location of the oldest branch on the list of predictions, the predicted target address must equal the next instruction address determined by executing the branch. If the two addresses are not the same, this is usually because the executed branch is not taken though it was predicted to be taken. In this case, the predicted branch has a nonsequential next address and the executed branch has a sequential next address. But, since the predicted target address is based on the address tag of the successor block, this comparison also detects that cache replacement has removed the original target entry, meaning that the wrong instructions are now in the target block. Finally, comparing target addresses checks that indirect branches were properly predicted, because the target addresses of indirect branches can change during execution.

If either of the foregoing conditions does not hold, the instruction fetcher has mispredicted a branch. The instruction fetcher uses the location of the branch determined by the execution unit to update the appropriate cache entry.

4.3.4 Hardware and Performance Costs

The principal hardware cost of using the instruction cache to predict branches is the increase in the cache size caused by the successor index and branch block index in each entry. For a 64-Kbyte, direct-mapped cache, these add about 11% to the cache storage. However, the performance increase obtained by branch prediction seems worth this cost, especially considering the typical incremental performance improvements that result in the scalar processor when the cache is doubled in size. As Figure 4-11 shows, branch prediction in the cache improves average performance by about 30% with a four-instruction decoder. Doubling the size of a 64-Kbyte cache in the scalar MIPS architecture yields a performance improvement of about 10-15% [Przybylski et al. 1988].

This scheme retains branch history only to the extent of tracking the most recent taken branches. Branch-prediction accuracy can be increased by retaining more branch-history information [Lee and Smith 1984], but this is significantly more expensive in the current proposal than in most other proposals. The current proposal reduces storage by closely associating branch-prediction information with a cache block: this scheme predicts only one taken branch per cache block and predicts non-taken branches by not storing any branch information with the block. This has the advantage that any number of nontaken branches can be predicted without contending with taken branches for cache entries. Figure 4-15 illustrates, with the set of bars labeled "predict all," the reduction in performance that would result if information were retained for every branch rather than only taken branches. The reduction in performance is due solely to contention for prediction entries between branches in the same cache block. This contention can be reduced by additional storage to hold branch information for more than one branch per block, but the slight improvement in prediction accuracy of more complex branch-prediction schemes (2-3% at best) is not worth the additional storage and complexity.

The requirement to update the cache entry when a branch is mispredicted conflicts with the requirement to fetch the correct branch target. Unless it is possible to read and write the fetch information for two different entries simultaneously, the updating of the fetch information on a mispredicted branch takes a cycle away from instruction fetching. Figure 4-16 indicates, with the set of bars labeled "update pred," the effect that this fetch stall has on performance for a two-instruction and a four-instruction decoder. For comparison, the results of the "pred" case from Figure 4-11 is repeated in Figure 4-16. The additional fetch stall causes only a small reduction in performance: 3% with a two-instruction decoder and 4% with a four-instruction decoder. This is reasonable, since the penalty for a mispredicted branch is large (see Figure 4-5) but is incurred infrequently; the additional cycle represents only a small proportional increase in the penalty (as with all simplifications we examine, this performance reduction is in relation to our standard processor model and does not assume any other incremental simplification).

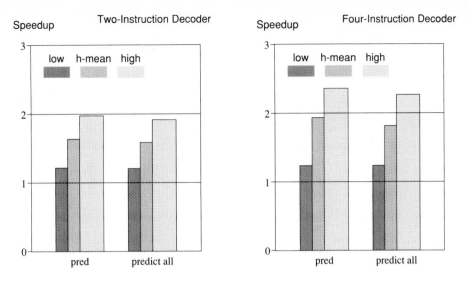

*Figure 4-15. Performance Decrease Caused by Storing
All Branch Predictions with Cache Blocks*

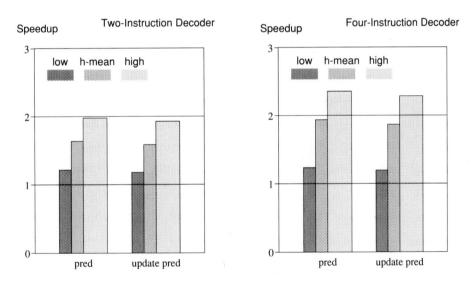

Figure 4-16. Performance Degradation Caused by a Single-Port Prediction Array

4.4 IMPLEMENTING A FOUR-INSTRUCTION DECODER

Section 4.2.4 showed that a four-instruction decoder yields higher performance
(20-25% higher) than a two-instruction decoder, because the four-instruction decoder is less sensitive to instruction alignment than the two-instruction decoder.

However, directly modeling a four-instruction decoder after a single-instruction decoder is not cost effective. In a straightforward implementation, decoding four instructions per cycle requires eight read ports on both the register file and the reorder buffer and eight buses for distributing operands. Furthermore, with this four-instruction decoder, the execution hardware described in Section 3.4 requires a total of 210 comparators for dependency checking (192 in the reorder buffer for associative lookup and 18 to check dependencies between decoded instructions). At the same time, the performance benefit of a four-instruction decoder is limited by other performance constraints. It hardly seems worthwhile to double the amount of hardware over a two-instruction decoder for a 20-25% performance benefit as shown in Figure 4-11 (which includes the assumption that cycle time is not affected).

Furthermore, most of the capability of a straightforward four-instruction decoder is wasted. Figure 4-17 shows the demand for register-file operands made by a four-instruction decoder during the execution of the sample benchmarks. This distribution was measured with the decode stage occupied by valid instructions on every cycle, so this is an upper bound (there are no branch-delay cycles, although there are misalignment penalties). There are several reasons that the register demand is so low:

- Not all decoded instructions access two registers.

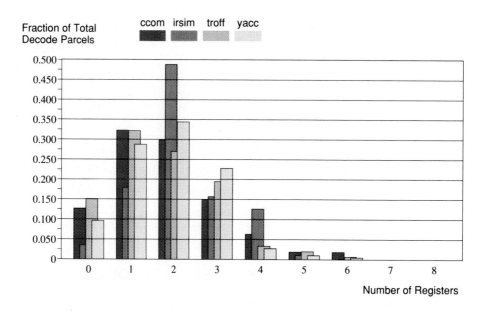

Figure 4-17. Register-Usage Distribution of a Four-Instruction Decoder–
No Branch Delays

- Not all decoded instructions are valid, because of misalignment.
- Some decoded instructions have dependencies on one or more simultaneously decoded instructions (the corresponding operands are obtained later by forwarding results directly to the reservation stations).

We can take advantage of the relatively low demand for register-file read ports if the decoder provides a limited number of read ports that are allocated, using arbitration, as other processor resources are. The amount of hardware in the reorder buffer also is reduced in this manner, since the reorder buffer has the same number of read ports as the register file. Figure 4-18 shows the results of constraining the number of registers available to a four-instruction decoder. This constraint reduces average performance by less than 2%.

Arbitrating for register-file ports requires that the decoder implement a prioritized select of the register identifiers to be applied to the register file. Instructions appearing in the leading decoder positions are allowed to access the register file before subsequent instructions, because they appear first in the instruction sequence. In RISC processors, the register file typically is accessed during the second half of the decode cycle, so the register identifiers must be valid at the register file by the midpoint of the decode cycle. For this reason, the prioritized register-identifier selection must be accomplished within about half of the processor cycle.

It is easy to design the instruction encoding so that register requirements are known early in the decode cycle–most RISC processors already have this feature.

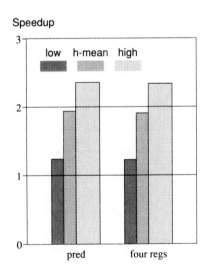

Figure 4-18. Performance Degradation Caused by Limiting a
Four-Instruction Decoder to Four Register-File Ports

However, it is somewhat more difficult to arbitrate register access among contending instructions. Table 4-1 shows an estimate of the amount of logic required to implement the prioritized register-identifier selection in two levels of logic. The eight register operands possibly required by a four-instruction decoder are shown across the top of Table 4-1. Each row of Table 4-1 shows, for each possible register operand, the number of gates and the number of inputs of each gate required to generate an enable signal. This signal selects the given register identifier at one input of an eight-to-one multiplexer, enabling the given operand to be accessed at the indicated register-file port. This logic follows the following general form:

- If the first register operand requires a register access, it is always enabled on the first port.
- If the second register operand requires a register access, it is enabled on the first port if the first register operand does not require an access, and is otherwise enabled on the second port.
- In general, the register identifier for a required access is enabled on the first port if no other previous operand uses a port, on the second port if one previous operand uses a port, on the third port if two previous operands use ports, on the fourth port if three previous operands use ports, and on no port (and the decoder stalls) if four previous operands use ports.

This type of arbitration occurs frequently in the superscalar processor to resolve contention for shared resources, though this is the first example we have seen. Resolving contention within two logic levels requires on the order of mn^3 gates, where m is the number of resources being contended for (in this case, four register-file read ports) and n is the number of contenders (in this case, eight register identifiers). A factor mn results from generating an enable per contender per resource, and a factor

Table 4-1. Estimate of Register-Port Arbiter Size: Two Logic Levels

Decode Reg #:	1	2	3	4	5	6	7	8
Enable to Port 0	—	1/1	1/2	1/3	1/4	1/5	1/6	1/7
Enable to Port 1		1/1	2/2	3/3	4/4	5/5	6/6	7/7
Enable to Port 2			1/2	3/3	6/4	10/5	15/6	21/7
Enable to Port 3				1/3	4/4	10/5	20/6	35/7
Totals:		2/1	4/2	8/3	15/4	26/5	42/6	64/7

Note: the first of each table entry is the number of gates required in the first level of logic, and the second is the number of inputs required by each gate. The sizes of the gates in the second logic level are indicated by the total number of gates in the first level.

n^2 results from the arbitration interaction between contenders. Since the largest first-level gate requires n minus 1 inputs, the amount of chip area taken by the first-level gates is on the order of mn^4. The size and irregularity of the prioritizer argue for implementing it with a logic array. However, regardless of the implementation, the prioritizer is likely to be slow.

The prioritizer is much smaller and more regular if it is implemented in serial logic stages. In this implementation, each register operand receives a port identifier from the preceding, adjacent operand. The operand either uses this port and passes the identifier for the next available port (or a disable) to its successor or does not use the port and passes the port identifier unmodified to its successor. The obvious difficulty with this approach is the number of levels required: the register-port arbiter requires approximately 14 logic levels using this approach. This number of logic levels is not out of the question, but makes it difficult to provide the register identifiers within a half cycle.

Thus, even though the number of register-file ports can be reduced, the complexity of the arbiter reduces the advantage somewhat. Still, the savings is significant because the register file and the reorder buffer are relatively large. Arbitration halves the size of these resources and increases their utilization.

4.5 IMPLEMENTING BRANCHES

The techniques for performing branches in a scalar RISC processor cannot be used directly in the superscalar processor. The superscalar processor may decode more than one branch per cycle, may have to retain a branch instruction several cycles before it is executed, and may have several unexecuted branches pending at any given time. In this section, we will consider the performance value of these complications and look at ways to simplify branch decoding and execution. To avoid including unrelated effects in the comparison between the superscalar and the scalar processors, the branch instructions and the pipeline timing of branches are assumed to be equivalent to those of the R2000 processor (many of the branch instructions in the R2000 perform comparison operations as well as branching, and these branches require a separate, but simple, functional unit to perform the comparisons; the comparisons cannot be performed by the ALU without conflicts).

4.5.1 Number of Pending Branches

Having a number of outstanding, unexecuted branches in the processor is the natural result of using branch prediction to overcome the branch delay. Between the time that a branch is decoded and its outcome is determined, one or more subsequent branches may be decoded. For best instruction throughput, subsequent branches should also be predicted and removed from the decoder. This not only avoids decoder stalls, but also provides additional instructions for issue in the likely event that subsequent branches are predicted correctly. However, the number of pending branches directly affects the size of the branch reservation station and the size of the branch-prediction list.

Figure 4-19 shows the reduction in performance as the allowed number of outstanding branches is decreased. The processor hardware used to obtain these results has six reservation-station entries for branches, rather than four, to accommodate all possible outstanding branches. With either a two-instruction or four-instruction decoder, nearly maximum performance is achieved by allowing up to four outstanding branches. In this case, the superscalar processor is issuing useful instructions from as many as five different instruction runs at once.

4.5.2 Order of Branch Execution

As long as branches are correctly predicted, branches can be executed in any order, and multiple branches can be executed per cycle. This improves instruction throughput and decreases the branch penalty by decreasing the chance that a mispredicted branch has to wait several cycles while previous, successfully predicted branches complete in sequence. Of course, if any branch is mispredicted, all subsequent results must be discarded even though subsequent branches appear to be correctly predicted.

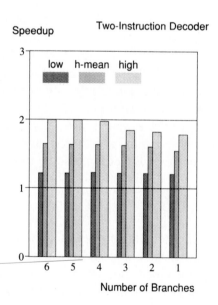
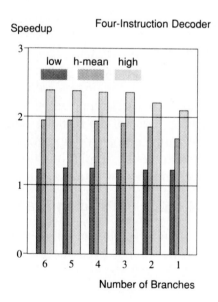

Figure 4-19. Reducing the Number of Outstanding Branches

Figure 4-20 shows the performance benefit of executing multiple, out-of-order branches per cycle. There is no increase in performance with a two-instruction decoder, because performance is limited by instruction fetching. And, although there is a slight increase in performance with a four-instruction decoder (about 3%), the increase is not large enough to warrant the additional hardware to find and execute multiple branches per cycle.

4.5.3 Simplifying Branch Decoding

Implementing a minimum branch decode-to-execution delay, and thus minimizing the penalty of a misprediction, requires that the branch-target address be determined during the decoding of a branch. The target address may be needed as soon as the following cycle, for detecting a misprediction and fetching the correct target instructions. The R2000 instruction set follows the common practice of computing branch-target addresses by adding program-counter-relative displacements to the current value of the program counter. Since the superscalar processor may decode more than one branch per cycle, computing potential branch-target addresses for all decoded instructions would require an adder per decoded instruction.

Fortunately, there is not a strong need to compute more than one branch-target address per decode cycle. As Figure 4-21 shows, there is only a slight performance decrease (about 2%) caused by imposing on the decoder a limit of one target address per cycle. The decrease is slight because the branch-prediction logic is limited to

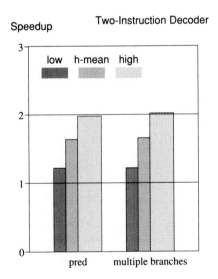
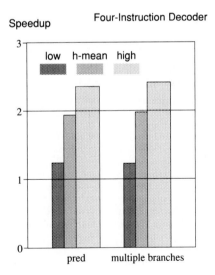

Figure 4-20. Performance Increase by Executing Multiple Correctly Predicted Branches per Cycle

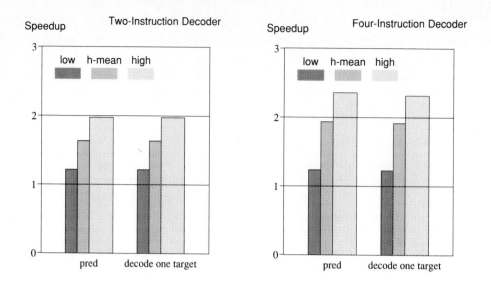

*Figure 4-21. Performance Decrease Caused by Computing
One Branch-Target Address per Decode Cycle*

predicting only one branch per cycle, and the execution hardware is limited to executing only one branch per cycle. Placing an additional limit in the decode stage may occasionally stall the decoding of a second branch and the instructions following this branch, but this limit is not severe compared to the other branch limits. This limitation can be removed entirely if the relative-address computation is performed during the execution of the branch rather than during decode, in which case all decoded branches are simply placed into the branch reservation station. But this adds a cycle to the best-case branch delay and increases the penalty of a mispredicted branch by one cycle, because the address is not available immediately after the decode cycle. The additional cycle of branch delay reduces performance by about 3% with a two-instruction decoder and by about 4% with a four-instruction decoder, so this is not a good alternative given that it is more complicated to place multiple branch instructions per cycle into the branch reservation station. Having a minimum branch delay is more important than decoding multiple branches per cycle.

Even if the superscalar processor has only one target-address adder, computing the target address still is more complex than in the scalar processor. Instructions in the decoder must arbitrate for the address-computation hardware, with the first branch instruction having priority. Also, the program-counter-relative computation must take into account the position of the branch instruction in the decoder, because the program counter of the branch instruction depends on its position in the decoder.

4.6 REDUCING THE PENALTY OF PROCEDURAL DEPENDENCIES: OBSERVATIONS

To sustain adequate instruction throughput, the instruction fetcher of a superscalar processor must be able to fetch successive runs of instructions without intervening delay, and the processor should have a wide (four-instruction) decoder. The instruction fetcher also should be able to fetch through as many as four runs of instructions before an unresolved branch is allowed to stall the decoder.

Hardware branch-prediction can be added to the instruction cache for a small relative cost. Alternatively, software branch-prediction incurs almost no hardware cost and has the added advantage that software can, based on the software prediction, align and merge instruction runs–except that this approach is not code compatible with existing processors. Although we did not examine software branch-prediction, aligning, and merging in detail, software branch-prediction is about as accurate as hardware branch-prediction, and we have seen indications that software aligning and merging can increase performance by about 5%. If software branch-prediction is used, the instruction fetcher still must be able to fetch instruction runs without intervening delays; the instruction-set architecture cannot have fully general relative branches, and branches must be readily identifiable by the fetch hardware (even having the decode hardware identify branches imposes too much of a penalty). Existing processors do not have such restrictions, so the resulting hardware simplifications are possible only with new, incompatible instruction sets.

A four-instruction decoder provides higher sustained instruction bandwidth than a two-instruction decoder and also allows a higher peak execution rate. The four-instruction decoder can be constructed without the eight register-file read ports that are implied by a simplistic implementation. A two-instruction decoder may be justified, however, because the hardware is simpler, though performance is lower. The decision between a two-instruction and a four-instruction decoder also can depend on the position of the instruction cache with respect to the processor. If the processor and cache are on separate chips, it may be too expensive to communicate four instructions in a single cycle. Throughout subsequent chapters, we will examine performance implications for processors both with two-instruction and with four-instruction decoders.

Since efficient instruction fetching relies on some sort of branch prediction, there must be a recovery mechanism for undoing the effects of instructions executed along a mispredicted path. This mechanism is the topic of the following chapter. Following this, we will examine the relationship of the recovery mechanism to register renaming.

Chapter 5

The Role of Exception Recovery

A superscalar processor achieves high instruction throughput by fetching and issuing instructions under the assumption that branches are correctly predicted and that exceptions do not occur. This is sometimes called *speculative execution*. Speculative execution allows instruction execution to proceed without waiting on the completion of previous instructions. However, the processor must produce correct results even when these assumptions fail. Producing correct results requires *recovery* and *restart* mechanisms. Recovery mechanisms cancel the effects of instructions that were issued under false assumptions, and restart mechanisms reestablish the correct instruction sequence.

There are an almost endless number of ways to restart a processor after an exception: this is one of the most complex areas of computer architecture. However, we have a very specific objective for the superscalar processor, and this objective limits our range of alternatives. In the superscalar processor, recovery and restart occur frequently, because of mispredicted branches, and must be accomplished rapidly. We cannot adopt schemes that depend on exceptions being infrequent, such as schemes that rely heavily on software intervention to reestablish correct operation. The recovery and restart mechanisms must be in hardware, even when used to recover from branches mispredicted by software.

In this chapter, we will consider the cost, complexity, and performance of various recovery and restart mechanisms. These mechanisms are essential to performance in a superscalar processor. Specifically, we will focus on hardware for implementing precise interrupts, because this hardware limits the damage that can be caused by incorrect operation and thus permits efficient recovery.

5.1 BUFFERING STATE INFORMATION FOR RESTART

To permit efficient recovery, we are interested in two sets of state information. The first set is the state required for computation; the second set is the state at the point of an exception, which contains no incorrect value resulting from the exception. In-order completion makes these two sets of state identical and makes it trivial to implement recovery, but in-order completion yields low performance. The implementation of precise interrupts with out-of-order completion requires buffering–that is, additional storage locations–so that the processor can maintain both sets of state at once.

This section describes the different types of state information to be maintained and describes four proposed buffering techniques that maintain the appropriate state information: *checkpoint repair*, a *history buffer*, a *reorder buffer*, and a *future file*. This section contrasts the four buffering techniques and identifies those techniques appropriate for restarting a superscalar processor.

We are concerned in this section with the state held in processor registers. Chapter 8 will explore techniques that can be used for state in external memory. External memory requires less sophisticated techniques than those described here, because the memory is written less frequently than processor registers.

5.1.1 In-Order, Lookahead, and Architectural State

To aid understanding of the mechanisms for precise interrupts, this section introduces the concepts of *in-order, lookahead,* and *architectural* state, illustrated in Figure 5-1. The processor illustrated in Figure 5-1 has out-of-order issue and out-of-order completion. Figure 5-1 shows the register assignments performed by a sequence of instructions; in this sequence, completed instructions are shown in boldface.

The *in-order state* is made up of the most recent assignments performed by the longest continuous sequence of completed instructions. In Figure 5-1, the assignments performed by three of the first four instructions are part of the in-order state (as are assignments performed by previous instructions that are not shown and assumed completed). The assignment to *R7* in the second instruction does not appear in the

Instruction Sequence	Items in In-Order State	Items in Lookahead State	Items in Architectural State
R3 := ... (1)	**R3** := ... (1)		
R7 := ... (2)			
R8 := ... (3)	**R8** := ... (3)		
R7 := ... (4)	**R7** := ... (4)		*R7* := ... (4)
R4 := ... (5)		*R4* := ... (5)	*R4* := ... (5)
R3 := ... (6)		*R3* := ... (6)	
R8 := ... (7)		*R8* := ... (7)	*R8* := ... (7)
R3 := ... (8)		*R3* := ... (8)	*R3* := ... (8)

Figure 5-1. Illustration of In-Order, Lookahead, and Architectural States

in-order state because it has been superseded by the assignment to $R7$ in the fourth instruction. Though the sixth instruction is shown completed, its assignment is not part of the in-order state because it has completed out of order: the fifth instruction has not yet completed.

A processor with in-order completion maintains the in-order state at all times in all storage locations–this is the primary value of in-order completion. With out-of-order completion, in contrast, we have additional sets of state. The *lookahead state* consists of all assignments, starting with the first uncompleted instruction, to the end of the sequence, shown by italicized assignments in Figure 5-1. The lookahead state is made up of actual register values as well as pending updates, since there are both completed and uncompleted instructions (in the hardware, pending updates are represented by tags). Because of possible exceptions, all pending assignments are retained in the lookahead state and no value is superseded. For example, the assignment to $R3$ in the sixth instruction is not superseded by the assignment to $R3$ in the eighth instruction; both assignments should be considered part of the lookahead state and added in sequence to the in-order state. To illustrate further how the lookahead state is added to the in-order state, the assignments of the fifth and sixth instructions will become part of the in-order state as soon as the fifth instruction completes successfully, and at that time the assignment to $R3$ by the sixth instruction will supersede the assignment to $R3$ by the first instruction in the in-order state.

The *architectural state* consists of the most recently completed and pending assignments to each register, relative to the end of the known instruction sequence, regardless of which instructions have been issued or completed. This is the state that must be accessed by an instruction following this sequence, for correct operation. In the architectural state, the pending assignment to $R3$ in the eighth instruction supersedes the completed assignment to $R3$ in the sixth instruction, because a subsequent instruction must get the most recent value assigned to $R3$. If a subsequent instruction accessing $R3$ were decoded before the eighth instruction completed, the decoded instruction would obtain a tag for the new value of $R3$, rather than an old value for $R3$. Note that the architectural state is not separate from the in-order and lookahead states but is obtained by combining these states.

All hardware implementations of precise interrupts described in the following sections must correctly maintain the in-order, lookahead, and architectural states. These implementations differ primarily in the mechanisms used to isolate and maintain these sets of state.

5.1.2 Checkpoint Repair

Hwu and Patt [1987] describe the use of *checkpoint repair* to recover from mispredicted branches and exceptions (see Figure 5-2). The processor provides a set of *logical spaces*, where each logical space consists of a full set of software-visible registers and memory. Of all the logical spaces, only one is used for current execu-

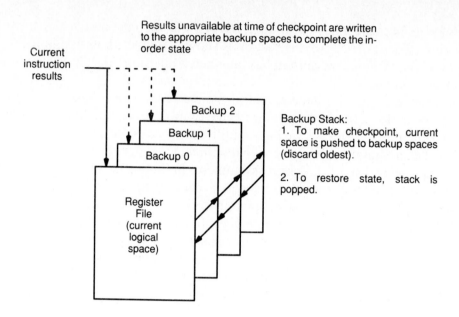

Figure 5-2. Checkpoint Repair

tion. The other logical spaces contain backup copies of the in-order state that correspond to previous points in execution. At various times during execution, a checkpoint is made by copying the architectural state (not the in-order state) of the current logical space to a backup space. Logical spaces are managed as a stack, so making a checkpoint discards the oldest checkpointed state. Since the copied state is not the in-order state at that time, the checkpointed state is updated as instructions complete, to bring it to the desired in-order state, and updating ceases as soon as the in-order state is reached. After the checkpoint is made, computation continues in the current logical space. If an exception occurs, all instructions preceding the point of exception are allowed to complete, if required, bringing the checkpointed state to the in-order state at the point of the exception. Restarting is accomplished, if required, by loading the contents of the appropriate backup logical space into the current logical space. The backup space used to recover depends on the location of the exception with respect to the location of the checkpoint in the instruction sequence.

Checkpoints are not made at every instruction, because the overhead would be too high. Hwu and Patt propose two checkpoint mechanisms: one for exceptions and one for mispredicted branches. This approach is based on differences between exception restart and misprediction restart and is intended to reduce the number of logical spaces. Exceptions can happen at any instruction in the sequence, but exception restart is required infrequently, so it is acceptable for exception restart to take a long time. Thus, checkpoints for exception restart are made infrequently and at widely separated points in the instruction sequence. If an exception does occur, the processor recovers the state at the point of exception by recovering the checkpointed state preceding the point of exception (this may be the state several instructions before the point of exception) and then executing instructions in order up to the point of exception. On the other hand, branch misprediction occurs only at branches, rather than at any instruction, and misprediction restart is required frequently. Thus, checkpoints for misprediction restart are made at every branch and contain the precise state to restart execution immediately after the mispredicted branch is detected.

If the processor were to spend much time making copies of state during checkpointing, performance would suffer because checkpointing is a frequent operation— at least as frequent as every branch. To avoid this overhead, Hwu and Patt propose implementing the logical spaces with multiple-bit storage cells. For example, each bit in the register file might be implemented by four bits of storage: one bit for the current logical space and three bits for three backup logical spaces. This complexity is the most serious disadvantage of this proposal [Uvieghara et al. 1990]. There is a tremendous amount of storage for the logical spaces, but the contents of these spaces differ by only a few locations, depending on the number of results produced between checkpoints. It is much more efficient to simply maintain these state differences in a dedicated structure such as the history buffer or reorder buffer described in the following sections (although we will also see that the reorder buffer is preferred over the history buffer).

5.1.3 History Buffer

Figure 5-3 shows the organization of a *history buffer*. The history buffer was proposed by Smith and Pleszkun [1985] as a means for implementing precise interrupts in a pipelined scalar processor with out-of-order completion. In this organization, the register file contains the architectural state, and the history buffer stores items of the in-order state which have been superseded by items of lookahead state (that is, it contains old values which have been replaced by new values; hence the name *history buffer*). The in-order state and lookahead state are not maintained explicitly, but either one can be recovered from the state in the register file and the history buffer.

The history buffer is managed as a last-in, first-out (LIFO) stack. When an instruction is decoded, the current value of the instruction's destination register is pushed onto the top of the history buffer. Other values in the history buffer are pushed down toward the bottom of the history buffer. The value already at the bottom of the history buffer is discarded if the associated instruction completed successfully (the associated instruction is the one that caused the value to be pushed onto the history buffer). If the associated instruction has not completed, instruction decoding is stalled and no values are pushed onto the history buffer until the instruction does complete.

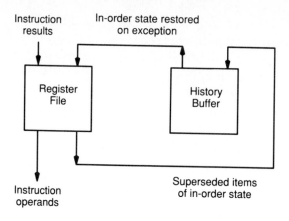

Figure 5-3. History Buffer Organization

When an instruction completes with an exception, the processor suspends instruction decoding and waits until all other pending instructions (in the window or in the processor pipelines) are completed. These instructions write their results into the register file. Following this, all values are popped from the top of the history buffer, one at a time, and written back into the appropriate locations of the register file, up to the point of the instruction with the exception. This restores the register file to the in-order state at the point of the exception. Values are popped in LIFO order so that, if there are multiple values from the same register, the oldest value will be placed into the register file last. Other, newer values are part of the lookahead state.

The history buffer has two significant disadvantages that are avoided by other schemes presented shortly. First, it requires additional ports on the register file for transferring the superseded values into the history buffer. For example, a four-instruction decoder requires four additional ports for reading all result-register values in one cycle. These additional ports contribute nothing to performance. Second, when an exception or branch misprediction occurs, the history buffer requires several cycles to restore the in-order state into the register file. These cycles are probably unimportant for exceptions, because exceptions are generally infrequent. However, the additional cycles are excessive for mispredicted branches. For these reasons, the history buffer is inappropriate for restart in a superscalar processor.

5.1.4 Reorder Buffer

Figure 5-4 shows the organization of a *reorder buffer* [Smith and Pleszkun 1985]. In this organization, the register file contains the in-order state, and the reorder buffer contains the lookahead state. The architectural state is obtained by combining the in-order and lookahead states, ignoring all but the most recent updates to each regis-

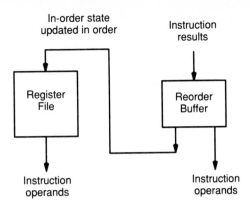

Figure 5-4. Reorder Buffer Organization

ter. The in-order and lookahead states can be combined using associative hardware as we have seen in our standard processor model [Sohi and Vajapeyam 1987]. Alternatively, software can unify these states if the reorder buffer is visible to software [Pleszkun et al. 1987], except that this software is not compatible with conventional processors.

The reorder buffer is managed as a first-in, first-out (FIFO) queue. When an instruction is decoded, it is allocated an entry at the top of the reorder buffer. The result value of the instruction is written into the allocated entry after the instruction completes. When the value reaches the bottom of the reorder buffer, it is written into the register file if there are no exceptions associated with the instruction. If the instruction is not complete when its entry reaches the bottom of the reorder buffer, the reorder buffer does not advance until the instruction is complete, but instruction decoding can proceed as long as there are available entries to be allocated. If there is an exception, the contents of the reorder buffer are discarded, and the processor reverts to accessing the in-order state in the register file.

The reorder buffer has the disadvantage of requiring associative lookup to combine the in-order and lookahead states (ignoring the possibility of software solutions). Furthermore, this associative lookup is not straightforward, because it must obtain the most recent assignment if there is more than one assignment to a given register. Finding the most recent assignment requires that the associative lookup be prioritized by instruction order and that the reorder buffer be implemented as a true FIFO array, rather than as a more simple, circularly addressed register array.

However, the reorder buffer overcomes all disadvantages of the history buffer. Unlike the history buffer, the reorder buffer does not require additional ports on the register file for reading superseded values. Upon exception, the history buffer takes several cycles to restore the in-order state, but the reorder buffer discards the lookahead state beyond the point of exception in a single cycle and thus permits instruction execution to restart immediately. Although the reorder buffer appears to have more ports than are required to supply instruction operands, the additional ports for writing results to the register file are simply the outputs of the final entries of the FIFO and do not have the costs associated with true ports (this consideration further argues for implementing the reorder buffer as a FIFO).

5.1.5 Future File

The associative lookup in the reorder buffer is avoided using a *future file* to contain the architectural state, as shown in Figure 5-5 (this future file is a slight variation of the future file proposed by Smith and Pleszkun [1985]). In this organization, the register file contains the in-order state and the reorder buffer contains the lookahead state. Both the reorder buffer and the register file operate as described in the preceding section, except that the architectural state is duplicated in the future file. The future file can be structured exactly as the register file is and can use identical access techniques.

With a future file, operands are not accessed from the reorder buffer. The reorder buffer only updates the in-order state to the register file. During instruction decode, register identifiers are applied to both the register file and the future file. If the future file has the most recent entry for a register, that entry is used (this entry may be a tag rather than a value if the entry has a pending update); otherwise, the value in the register file is used (values in the register file are always valid, and there are no tags). Once a register in the future file has been marked as containing the most recent value (or a tag for this value), subsequent instructions accessing this register obtain the

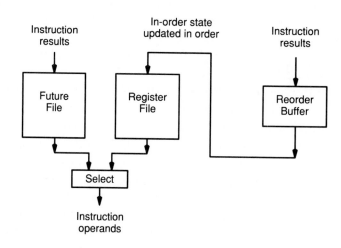

Figure 5-5. Future File Organization

value in the future file (or a tag in the future file). If an exception occurs, the processor recovers by completing all updates to the in-order state in the register file, using the reorder buffer, then discarding the contents of the future file and any remaining entries in the reorder buffer. After recovery, the processor reverts to accessing the in-order state in the register file.

When an instruction completes, its result value is written to the future-file location associated with the result register, unless this value has been superseded by the result of a subsequent instruction, in which case the value is simply discarded. In any case, the value is also written into the reorder-buffer entry that was allocated during decode. At this point, the result value in the future file has been marked as the most recent value, causing it effectively to replace the value in the register file. Once an entry in the future file is marked as most recent, it remains the entry used by subsequent instructions until the next exception occurs. The future file still may contain a tag instead of a value at various times during execution.

The future file described by Smith and Pleszkun [1985] is slightly different than the future file described here, because, in their organization, only the future file provides operands. Their scheme uses the register file to communicate the in-order state to an exception handler and requires that the in-order state be copied to the future file before restart. This is adequate to handle exceptions but is too slow for mispredicted branches. The scheme described here allows both the future file and the register file to provide operands and permits quick restart after a mispredicted branch, because the register file can supply all operands after recovery without copying values to the future file.

The future file overcomes the disadvantages of the history buffer without the associative comparators of the reorder buffer. However, this advantage is at the expense of an additional array (the future file) and the validity and tag bits associated with entries of this array.

5.2   RESTART IMPLEMENTATION AND EFFECT ON PERFORMANCE

Given that the processor contains adequate buffering to recover the in-order state at a previous point in execution, it still must be able to identify this state precisely. To restart execution at a previous version of the processor state, the processor must associate the exception with the proper version of state and must discard all updates following the point of exception without disturbing any state preceding the point of exception. For best performance, the processor hardware should restart automatically after detecting a mispredicted branch to avoid excessive penalties for misprediction. Other exceptions occur less often and probably require software intervention because the processor does not know how to recover. To allow restart after an exception, the processor need simply communicate a consistent, in-order state to a software exception handler.

5.2.1 Mispredicted Branches

A reorder buffer, used either with or without a future file, provides a straightforward mechanism for the processor to discard the state changes made after a mispredicted branch [Sohi and Vajapeyam 1987]. However, out-of-order issue and completion make it somewhat difficult for the processor to distinguish instructions following the branch–whose updates should be discarded–from instructions preceding the branch–whose updates should not be discarded. When the mispredicted branch is detected, instructions of either type can be in various stages of execution. Figure 5-6 shows the action required to correct the processor state if a mispredicted branch follows the sixth instruction in the sequence of Figure 5-1. Processor state must be recovered at the point of the mispredicted branch, but it is not correct to discard the entire lookahead state, because some of this state is associated with incomplete instructions which preceded the branch. These items in the lookahead state must be preserved. Furthermore, each previously superseded value in the lookahead state must be "uncovered" so that subsequent instructions along the correct path do not obtain operand values from instructions along the incorrect path. In this respect, the future file has a significant implication on the processor design, as discussed soon.

If a reorder buffer is used without a future file, the architectural state is provided via associative lookup in the reorder buffer. In this case, backing up the state after a mispredicted branch is simply a matter of clearing all entries that were allocated after

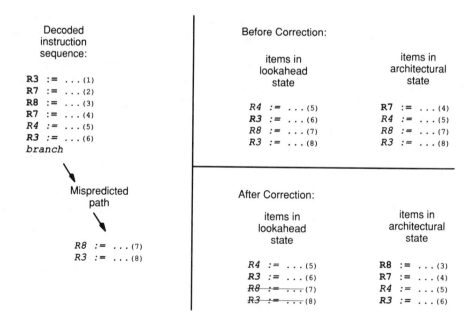

Figure 5-6. Correcting State After a Mispredicted Branch

the mispredicted branch. The associative lookup is already prioritized so that the reorder buffer provides the most recent value; clearing the entries allocated following a mispredicted branch simply takes these incorrect entries out of consideration as being the most recent values. Clearing these entries requires the capability to reset a variable number of reorder-buffer entries at a variable starting point (that is, starting at the point of the mispredicted branch), but is otherwise not complex to implement.

In contrast, the future file stores the architectural state up to the current point of execution and does not distinguish correct from incorrect updates. If an incorrect value supersedes a correct value, the future file simply writes over the correct value and provides no means to recover this correct value. The processor cannot selectively discard incorrect values in the future file: it can discard incorrect values only by discarding all values in the future file. Before discarding all values in the future file, the processor must have the correct in-order state, at the point of the branch, in the register file. Otherwise, there is no way to recover and restart. The processor can therefore recover from a mispredicted branch only by waiting until the reorder buffer has made all the required updates to the in-order state in the register file, and must therefore wait until all instructions preceding the branch have been completed before it can restart after a mispredicted branch. This additional waiting adds to the branch delay

Figure 5-7 shows, for the sample benchmarks, the distribution of additional branch delay caused by waiting for the completion of all instructions preceding a mispredicted branch before restart. On the average, waiting adds no delay for about half of the mispredicted branches. The reason for this is that the outcome of a conditional branch depends on the results of other instructions, and, because of true dependencies, a branch is often the last or nearly last instruction completed relative to the point where the mispredicted branch is detected. It is likely that instructions preceding the branch point will have completed or be near completion by the time the outcome of the branch can be determined, so that there is only a small average penalty caused by waiting on these instructions. Figure 5-8 shows the impact of this additional penalty on performance. Adding this small penalty to an already large branch penalty (Figure 4-5, page 62) causes only a small proportional decrease in performance (4-5%).

The technique of waiting for the completion of preceding instructions before recovering from a mispredicted branch also can simplify the design of a reorder buffer, because it eliminates the need to clear a variable number of entries in the reorder buffer upon detection of a mispredicted branch. The processor simply waits until the appropriate in-order state is in the register file and then discards all entries in the reorder buffer. The performance penalty is the same as the penalty caused by using this technique with the future file.

Regardless of the mechanism used to restart after a mispredicted branch, there must be some mechanism for identifying the location of a branch point in the reorder

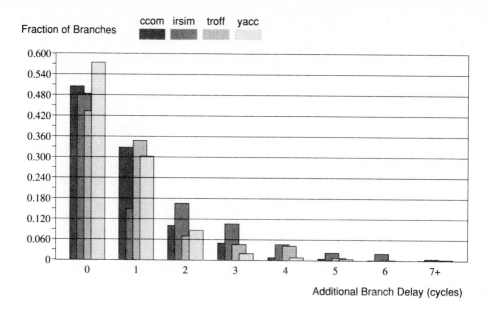

Figure 5-7. Distribution of Additional Branch Delay Caused by Waiting for the Completion of Instructions Preceding a Mispredicted Branch

buffer. Otherwise, the processor does not know which entries to reset or does not know when to discard the contents of the future file. The obvious way to identify this location is to allocate a reorder-buffer entry for each branch, even if the branch does not produce a result (a procedure call does write a return address into a register, but most branches do not write results). When the processor detects a mispredicted branch, it clears all reorder-buffer entries subsequent to the entry allocated for the branch, or, alternatively, it waits for this entry to reach the bottom of the reorder buffer before discarding the values in the future file and/or reorder buffer.

5.2.2 Exceptions

In general, exceptions occur because the processor hardware is not able to perform an operation correctly. The exception may indicate an error or may simply indicate that software intervention is needed to provide correct operation. In either case, the processor simplifies exception handling if it communicates a consistent in-order state to the exception handler. The alternative–providing bits of state and indications of the exception at various locations in the processor and system–is ad hoc and makes the exception-handling software very dependent on the processor implementation. The only advantage to the latter approach is that it yields simpler hardware, and many designers consider it acceptable to trade simpler hardware for more complex exception-handling software.

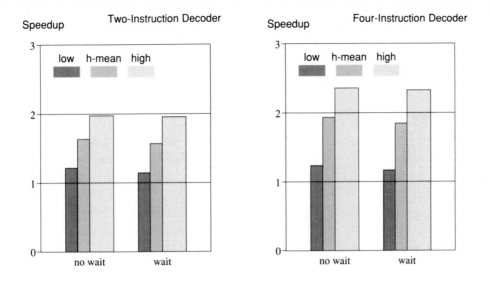

*Figure 5-8. Performance Degradation Caused by Waiting for the
Completion of Instructions Preceding a Mispredicted Branch*

However, we have already added hardware to the superscalar processor to re-cover efficiently from mispredicted branches, and it is an easy matter to use this hardware to provide precise exceptions. We encounter only a few complications be-yond those encountered in restarting after mispredicted branches. First, an excep-tion can happen at any instruction, in general. When recovery and/or restart occurs at any instruction, there must be a way to recover a program counter–in the case of mispredicted branches, the correct program counter is obtained when the branch is executed. Second, there must be a way to determine the in-order state at the point of the exception–in the case of mispredicted branches, this point is marked simply by allocating a reorder-buffer entry to each branch instruction.

To allow a program counter to be easily recovered upon exception, Smith and Pleszkun [1985] proposed keeping the program counter value of each instruction (determined during decode) in the reorder buffer along with the instruction result. This approximately doubles the storage in the reorder buffer. However, their pro-posal did not allocate reorder-buffer entries for branch instructions, because they were not concerned with restarting after mispredicted branches. In our case, the fact that the processor allocates entries for branches provides a convenient way to re-cover the program counter. This is accomplished by setting a program-counter reg-ister as every branch is removed from the reorder buffer and incrementing the value as nonbranching instructions are removed from the reorder buffer (the program counter is incremented by an appropriate amount, depending on the number of instructions removed). If an excepting instruction appears at the bottom of the reorder buffer, the program-counter register indicates the location of this instruction in memory. Maintaining a correct program counter requires that all instructions be allocated reorder-buffer entries, rather than just those that write registers, so that the processor knows the amount by which to increment the program counter. This is one motivation for allocating a reorder-buffer entry for each instruction; others are presented in the paragraphs that follow.

Determining the location of exceptions in the instruction stream requires that the reorder buffer be augmented with a few bits of instruction state to indicate whether or not the associated instruction completed successfully. For uniformity, this state information can also be used to indicate the location of branch points in the instruction stream. Indicating exceptions in this manner requires that each instruction be allocated a reorder-buffer entry, whether or not the instruction writes a processor register. An instruction may create an exception even though it does not write a register (the most common operation of this type is a store to the external memory). Furthermore, because exceptions are independent of register results, the instruction-state information in the reorder buffer must be written independently of other results.

It is simpler and faster for the decoder to allocate a reorder-buffer entry for every instruction than to allocate an entry just for those instructions which produce results. For the purpose of allocating the reorder buffer, the decoder need only determine which instructions are valid (as determined during instruction fetch), rather than also determine which instructions generate results.

An additional advantage to allocating a reorder-buffer entry for every instruction is that it provides a convenient mechanism for releasing stores to the data cache so that the in-order state is preserved in memory [Smith and Pleszkun 1985]. By keeping identifiers for the store buffer in the reorder buffer, the reorder buffer can signal the release of store-buffer entries as the corresponding stores reach the bottom of the reorder buffer. If the store buffer is allocated during the decode stage, store-buffer identifiers are readily available to be placed into the reorder buffer. Note, however, that a store exception occurring after a store has been released in this manner, for example a bus error, is not restartable unless there is some other mechanism to allow restart, because, after the store is released, the in-order state progresses beyond the point of the store. We could eliminate this problem by holding the release of the store until the store was fully complete with no exceptions, but, as we will see later, this has unacceptable performance because it limits the data throughput of the reorder buffer.

The essential advantage of allocating a reorder-buffer entry for every instruction is that it preserves instruction-order information in one location, after instruction decoding, even with out-of-order issue. This information can be useful for a variety of purposes, as we will see in Chapter 7.

5.2.3 The Effect of Recovery Hardware on Performance

Because the reorder buffer is of finite size, it can reduce performance. This is true whether or not the reorder buffer is used with a future file (however, the future file, by eliminating the associative lookup in the reorder buffer, may allow the reorder buffer to be larger). Instruction decoding must stall when there is no available reorder-buffer entry to receive the result of a decoded instruction. To avoid decoder stalls, the reorder buffer should be large enough to accept all instructions during the expected decode-to-completion period of most instructions.

Figure 5-9 shows the effect of reorder-buffer size on performance with a two-instruction decoder and a four-instruction decoder, respectively. The standard simulation model for the superscalar processor has two reorder buffers, to mimic the implementation of the R2000 processor and R2010 floating-point unit–one for integer results (and for other instructions, such as branches, that do not have results) and one for floating-point results. Only the size of the integer reorder buffer was varied for these results. The floating-point reorder buffer was held constant at eight entries.

There is little reduction in performance with 12 or more reorder-buffer entries, but performance decreases markedly with smaller numbers of entries. Considering instructions that generate results, the size of the reorder buffer is determined by the delays in generating results: it is necessary to provide enough buffering that the de-

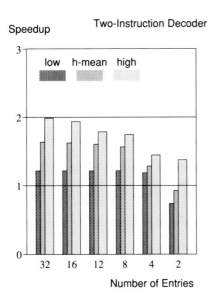 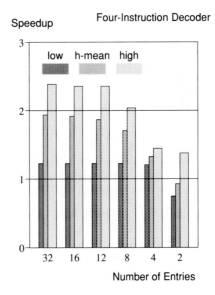

Figure 5-9. Effect of Reorder-Buffer Size on Performance:
Allocating for Every Instruction

coder does not stall if a result is not yet computed when the corresponding reorder-buffer entry reaches the bottom of the reorder buffer. When the reorder buffer is allocated for instructions without results (primarily stores and branches), the incremental demand placed on the reorder buffer by these instructions is directly proportional to their frequency relative to the result-producing instructions. The size of the reorder buffer that is determined by the peak demand of result-producing instructions is still adequate if the reorder buffer is allocated for all instructions. When the peak demand of result-producing instructions is achieved, there is, by implication, a local reduction in the number of instructions without results, and vice versa.

Figure 5-9 also shows that, if the reorder buffer is too small, performance of the superscalar processor can be less than the performance of the scalar processor. The reorder buffer causes the decode stage to depend on the writeback of results into the register file and prevents the processor pipeline from operating smoothly when the reorder buffer becomes full. When a group of instructions is decoded, the following group must wait until the first instructions have executed, written their results into the reorder buffer, and are about to have their results written into the register file. The scalar processor does not suffer the additional delay of writing results into the reorder buffer because the scalar processor model used for these results does not implement precise interrupts, even though it completes instructions out of order.

5.3   PROCESSOR RESTART: OBSERVATIONS

The best implementation of hardware recovery and restart has between a 12-entry and a 16-entry reorder buffer, with every instruction being allocated an entry in this buffer. A future file provides a way to avoid associative lookup in the reorder buffer, at the cost of an additional register array–a cost which can be prohibitive in architectures with a large number of registers. The future file also adds delay penalties to mispredicted branches, because it requires that the processor state be updated to the point of the mispredicted branch before the processor resumes instruction execution.

Instruction restart is an integral part of the superscalar processor, because it allows instruction decoding to proceed at a high rate. Even the software approaches to improving the fetch bandwidth proposed in Section 4.2, such as software branch-prediction, require nullifying the effects of instructions following a mispredicted branch. Furthermore, there is little sense in designing a superscalar processor with in-order completion, because of the detrimental effect on machine parallelism. Restarting a processor with out-of-order completion requires at least some of the hardware described in this chapter. If out-of-order issue is implemented, the reorder buffer provides a very useful means for recovering the instruction order that is abandoned after decode; for example, it helps in performing stores so as to preserve the in-order state in the external memory. As we will see in the next chapter, the reorder buffer is attractive also because it simplifies the implementation of register renaming.

Chapter 6

Register Dataflow

An instruction cannot be executed unless all of its input operands are available. The input operands to some instructions are available as soon as the program begins, but most operands are computed by instructions for use by other instructions. Sustaining a high rate of instruction execution thus requires a high rate of operand transfer between separate instructions.

Applications with data parallelism place few requirements on the communication delay between separate computations. Performance is not affected very much by the delay between the production of data and its use (after startup), because performance is determined principally by the rate at which results are produced (see Figure 6-1a). To permit a given computation to proceed at a given rate, input operands to the computation need only show up at the required rate. This is the essence of data parallelism. In contrast, the nature of the dependencies between computations in other applications make it impossible to separate the communication delay from the rate of computation. For example (see Figure 6-1b), if data produced by computation A is used by computation B to generate the next item required by A to produce the next item required for B, the rate of computation is strongly determined by the communication delay between A and B. This is the essence of instruction parallelism.

As one might expect, the term "instruction parallelism" does not so much denote a property of programs–though some programs do have more instruction parallelism than others–as it denotes the lack of data parallelism and, correspondingly, the lack of the performance potential that data parallelism permits. Exploiting instruction parallelism leads to the requirement of efficient, random communication between computations in the hope of extracting some measure of performance in the

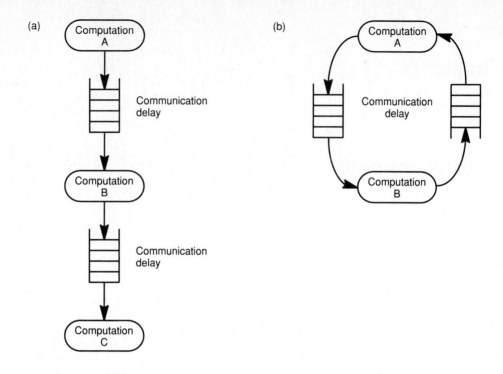

Figure 6-1. Computation Paradigms Where (a) Communication Delay Does Not Affect Computation Rate, and (b) Communication Delay Does Affect Computation Rate

face of complex, intertwined dependencies between the computations. Achieving performance on general-purpose applications requires multiple operand buses, result buses, and bypass paths to forward results directly to waiting instructions. Moreover, the processor requires low-delay mechanisms to control the routing of operand values to instructions and to insure that instructions are not issued until all input operands are valid. The most efficient processor organization is like that of a dataflow processor in many respects [Dennis and Misunas 1975] because, for example, after an instruction has been decoded, its execution is triggered by the availability, or arrival, of its input operands. Hence the term "dataflow" in the title of this chapter.

This chapter explores the flow of register-based data in a superscalar processor (Chapter 8 will consider memory-based data after we have had a chance to examine the implementation of out-of-order issue). We will consider two major topics in this chapter. The first topic is the detection of data dependencies that control the routing of data, including the detection or removal of storage conflicts that impede the flow of data. The second topic is the organization and control of the buses between the various processor units. As in previous chapters, we will examine the complexity of hardware and the performance hardware provides. Unfortunately, register dataflow is inherently complex, because the data dependencies between instructions are complex. In the superscalar processor, supplying operands for instructions is further complicated by the fact that the processor, on every cycle, can be generating multiple, out-of-order results and attempting to prepare multiple instructions for issue.

6.1 DEPENDENCY MECHANISMS

There is a bit of confusion surrounding the techniques for detecting and enforcing dependencies between instructions, in part due to the inherent complexity of this topic. Before launching into a detailed discussion of the various techniques, we will spend a few moments examining the subject at an abstract level. We will partition the subject of dependency checking to develop a framework for discussing dependency mechanisms.

The most simple hardware implementation relies on software to enforce dependencies between instructions. A compiler or other code generator can arrange the order of instructions so that the hardware cannot possibly see an instruction until it is free of true dependencies and storage conflicts. Unfortunately, this approach runs into several problems. Software does not always know the latency of processor operations, and so cannot always know how to arrange instructions to avoid dependencies. A commonly cited example of this problem arises because of the variable latency of data-cache accesses. An access may be a few cycles long, if data is found in the cache, or many cycles long, if data is not in the cache. Software cannot possibly know the latency of a cache access and cannot possibly satisfy all dependencies without some amount of hardware support (for example, hardware to suspend all execution when a cache miss is detected). In addition, branches create many possible execution paths through the code. When two paths converge, software must consider the potential dependencies along either path and satisfy the worst case along either path. This may slow the execution of a frequently executed path because of a dependency along a rarely executed path. Finally, there is the question of how the software prevents the hardware from seeing an instruction until it is free of dependencies. In a scalar processor with low operation latencies, software can insert *no-ops* in the code to satisfy data dependencies without too much overhead (the software reorganizer for the R2000 does this [Hennessy and Gross 1983]). If the processor is attempting to fetch several instructions per cycle, or if some operations take several cycles to complete, the number of *no-ops* required to prevent the processor from seeing dependent instructions rapidly becomes excessive, causing an unacceptable increase in code size. The *no-ops* use a precious resource–the instruction cache–to encode dependencies between instructions.

At some level, however, the foregoing arguments against having software enforce dependencies are rather weak. None of these problems is technically intractable, and, from an academic standpoint, it is arguable whether or not they are really very important. There are, however, more fundamental considerations. The first is that, when a processor permits out-of-order issue, it is not at all clear what mechanism software should use to enforce dependencies–software has little control over the behavior of the processor, so it is hard to see how software prevents the processor from decoding dependent instructions. The second consideration is that no existing binary code for any scalar processor enforces the dependencies in a superscalar processor, because the mode of execution is very different in the superscalar processor. Relying on software to enforce dependencies requires that the code be regenerated for the superscalar processor. Finally, the dependencies in the code are directly determined by the latencies in the hardware, so that the best code for each version of a superscalar processor depends on the implementation of that version. Code for higher-performance implementations, that is, implementations having smaller operation latencies, cannot both be compatible with lower-performance implementations *and* take advantage of the smaller operation latencies. A decision to use software to enforce dependencies is necessarily a decision against out-of-order issue and against binary code compatibility between implementations. Out-of-order issue and code compatibility require that hardware enforce dependencies. The fact that software techniques impose somewhat unusual requirements on the system–but yield simple hardware–makes the question of software versus hardware techniques one that can be debated endlessly. We cannot resolve this debate here–we will devote several of the following chapters to various forms of this question. For now, it should be possible to see that there is at least some motivation in favor of hardware dependency techniques.

On the other hand, there is at least some motivation against hardware dependency techniques, because they are inherently complex (software dependency-checking is also complex, but software is more capable than hardware in dealing with this complexity). If we assume instructions with two input operands and one output value, as holds for typical RISC instructions, then there are five possible dependencies between any two instructions: two true dependencies, two antidependencies, and one output dependency. Furthermore, the number of dependencies between a group of instructions–such as a group of instructions in a window–varies with the square of the number of instructions in the group, because each instruction must be considered against every other instruction. Complexity is further multiplied by the number of instructions that the processor attempts to decode, issue, and complete in a single cycle, because these actions introduce dependencies, are controlled by dependencies, and remove dependencies from consideration. The only aid we have in reducing complexity is that the dependencies can be determined incrementally, over many cycles, and this helps reduce the scope and complexity of the dependency hardware. There is, however, a fundamental complexity that we cannot remove.

In this section, we take care to separate hardware dependency *interlocks* from register renaming. An interlock prevents an instruction from being executed until the instruction is free of dependencies. In a strict sense, interlocks are needed only to enforce true dependencies, although they can be used also to enforce anti- and output dependencies. Renaming hardware removes anti- and output dependencies altogether, and thus is an alternative to interlock hardware for these storage conflicts. The attractiveness of register renaming is determined in comparison to other interlock mechanisms that enforce storage conflicts. For this reason, it is useful to examine the cost, complexity, and performance of various implementations of renaming in comparison to other interlock mechanisms.

This section describes the advantages of renaming and shows how renaming is implemented with a reorder buffer and a future file. This section also evaluates other interlock mechanisms that enforce anti- and output dependencies. We will focus on measuring the relative importance of various hardware features and identify the primitive hardware operations that enforce dependencies, in an attempt to identify areas where the hardware can be simplified or where performance can be improved. This approach suggests a new alternative to renaming, here called *partial renaming*, that has nearly all of the performance of renaming. Despite this new alternative, this section argues that–if dependency hardware is required–renaming with a reorder buffer is the most desirable alternative. The question of whether or not dependency hardware is required is somewhat controversial–this is a topic that we will return to several times throughout the rest of this book.

6.1.1 The Value of Register Renaming

To gauge the value of register renaming, we need something to compare against. The simplest alternative to renaming stalls instruction decoding when a dependent instruction is decoded and resumes instruction decoding when the dependency is removed. However, this is just in-order issue. We have already considered the performance of in-order issue and are more interested in a technique that supports out-of-order issue, but without renaming.

A simple alternative to renaming is based on Thorton's [1970] register *scoreboard* used in the CDC6600. In particular, we will use a variant of Thorton's algorithm proposed by Weiss and Smith [1984]. Weiss and Smith's variant of Thorton's algorithm uses a single bit associated with each register (the scoreboard bit) to indicate that a register has a pending update. When an instruction is decoded that will write a register, the processor sets the associated scoreboard bit. The scoreboard bit is reset when the write actually occurs. Since there is only one scoreboard bit indicating whether or not there is a pending update, there can be only one such update for each register. The scoreboard stalls instruction decoding if a decoded instruction will update a register that already has a pending update (indicated by the scoreboard bit being set). This avoids output dependencies by allowing only one pending update to a register at any given time.

When an instruction is decoded, the processor checks the scoreboard bit of the instruction's operand registers. If a scoreboard bit is set, the instruction has a true dependency on an uncomputed result. Instead of stalling the instruction decoder in this case (which would cause in-order issue), Thorton's algorithm places the dependent instruction into an instruction window (Thorton used the input registers at the functional units to implement a limited window, but Weiss and Smith used reservation stations as in our standard processor model). Copies of those input operands that are valid are also placed into the instruction window along with the instruction, avoiding antidependencies because these copies cannot be overwritten by subsequent instructions. For those operands that are not valid yet (those that have the scoreboard bit set), the processor places the corresponding register identifiers in the instruction window. When an instruction completes, its result-register identifier is compared to the identifiers in the window, and the result value is placed in the window for each instruction that is waiting on the value. This removes true dependencies from one or more instructions, possibly releasing some instructions for issue. Note that the register identifiers serve much the same purpose as result tags do in our standard processor model.

The term "scoreboarding" as we are using it is quite different from the term as it is applied to the Motorola 88100 processor [Motorola 1989] and the Intel 80960CA processor [Intel 1989b]. These processors use a hardware structure similar to Thorton's scoreboard but issue instructions in order rather than out of order. The term "scoreboarding" historically has been used in the context of allowing instruction decoding to progress, using some sort of instruction window, even after the processor encounters a true dependency. The 88100 and 80960CA do not fit this definition, because they stall instruction decoding when a true dependency or storage conflict is detected, though these processors do use scoreboards to detect these dependencies and conflicts. Since the 88100 and 80960CA scoreboards essentially detect and enforce dependencies for in-order issue, they should not be considered representative of scoreboarding as we are using the term here. We are examining dependency mechanisms in the context of out-of-order issue and so use the term "scoreboarding" in its historical sense.

Because it is a single bit of information, the scoreboard bit is easy to manage and to use. Register renaming, in contrast, uses multiple-bit tags to identify the various uncomputed values, some of which values may be destined for the same processor register (that is, the same program-visible register). Renaming requires hardware to allocate tags from a pool of available tags that are not currently associated with any value and requires hardware to free the tags to the pool once the values have been computed. Furthermore, since scoreboarding allows only one pending update to a given register, the processor is not concerned about which update is the most recent. This further simplifies the hardware in comparison to renaming, because renaming allows more than one pending update and requires that the processor be aware of the most recent update and not allow older updates to overwrite the most recent value. In theory, scoreboarding also does not require additional hardware registers to be renamed, although we will see later that this is not a very strong objection, because the reorder buffer, already included to permit recovery from mispredicted branches, provides this storage without much additional cost.

Having briefly described scoreboarding as an alternative to renaming, we now return to the original question of the value of register renaming. Scoreboarding is the simplest hardware technique that supports out-of-order issue, so we express the value of renaming in terms of its performance relative to scoreboarding. Figure 6-2 shows the performance of our standard superscalar processor with scoreboarding; the performance of renaming is shown here for comparison. Scoreboarding reduces average performance by about 15% with a two-instruction decoder and by about 21% with a four-instruction decoder. This is a little surprising, because scoreboarding and renaming differ only in one respect: renaming allows multiple instances of registers whereas scoreboarding does not. In terms of their efficiency with respect to true and antidependencies, scoreboarding and renaming are equivalent. It is apparent that the ability to create multiple instances of registers is important. However, these results are based on code generated by a compiler that performs aggressive register allocation and that often reuses the same registers for different values. This

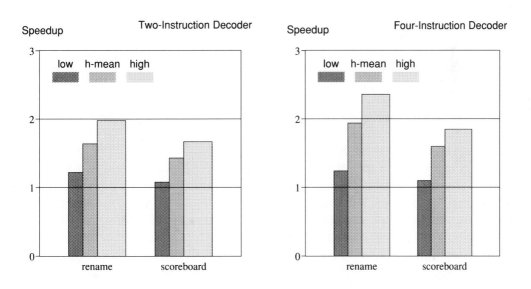

Figure 6-2. Performance of Scoreboarding Compared to Renaming

might exaggerate the advantage of register renaming. An architecture with a larger number of registers and different compiler technology would not experience the same advantage, but the degree to which this is true is very difficult to quantify.

The following two sections examine two implementations of renaming. We will assume that a reorder buffer or future file is already present in the implementation to allow efficient recovery from mispredicted branches, and will examine what hardware is required to implement renaming with these structures. This helps illuminate the cost and complexity of renaming before we move on to identify other alternatives to either renaming or scoreboarding. Ideally, we would like to find an alternative that is about as simple as scoreboarding but has about the same performance as renaming.

6.1.2 Register Renaming with a Reorder Buffer

A reorder buffer that uses associative lookup to form the architectural state (Section 5.1.4) provides a straightforward implementation of register renaming. The associative lookup maps the register identifier to the reorder-buffer entry as soon as the entry is allocated, and, to avoid output dependencies, the lookup is prioritized so that only the value for the most recent assignment is obtained if the register is assigned more than once (a tag for this value is obtained if the result is not yet available). There can be as many instances of a given register as there are reorder-buffer entries, so there are no storage conflicts between instructions. The values for the different instances are written from the reorder buffer to the register file in sequential order. When the value for the final instance is written to the register file, the reorder buffer no longer maps the register; the register file contains the only instance of the register, and this is the most recent instance. The only cost that renaming adds to the reorder buffer, over that required for recovery, is the logic to allocate and free the tags. The tags enforce dependencies within the lookahead state, even for values that may have been superseded in the architectural state. The tags are not required for recovery.

6.1.3 Renaming with a Future File: Tomasulo's Algorithm

Renaming with a reorder buffer relies very much on the associative lookup in the reorder buffer to map register identifiers to values. If we eliminate this associative lookup using a future file, we are faced with a new problem. In the reorder buffer, the associative lookup is prioritized so that the reorder buffer always provides the most recent value in the register of interest (or a tag). The reorder buffer also writes values to the register file in order, so that, if the value is not in the reorder buffer, the register file must contain the most recent value. The future file does not have these properties: a value presented to the future file to be written may not be the most recent value destined for the corresponding register, and the value cannot be treated as the most recent value unless it actually is. The future file must therefore keep track of the most recent update and check that each write corresponds to the most recent update before it actually performs the write.

To solve this problem, we can adapt Tomasulo's algorithm [Tomasulo 1967] to the future file. We are already using many components of this algorithm in our standard model, but we have omitted one key component that is not required with a reorder buffer. Tomasulo's algorithm implements renaming by associating a tag with each register. This tag identifies the most recent value to be assigned to the register and serves to rename the corresponding register (in Tomasulo's implementation, the tag was the identifier of the reservation station containing the assigning instruction, but, as Weiss and Smith [1984] point out, the tag can be any unique identifier). If there is no pending update to the register, the tag is ignored. When an instruction is decoded, and the instruction writes a register, an unused tag (selected by any number of means) is written into the tag entry corresponding to the result register. The tag identifies the result value that will be produced by the instruction. The tag is written regardless of any other pending updates to the register, because the currently decoded instruction performs the most recent update. When an instruction completes, its result value is written into the register file only if the tag for the result value matches the tag associated with the register. This avoids output dependencies while allowing multiple pending updates: if the result tag and the register tag do not match, the result does not correspond to the most recent pending update to the register.

Because Tomasulo's algorithm in its original form allows new values to overwrite older values in the in-order and lookahead states, it does not permit precise interrupts. However, Tomsulo's algorithm is easily adapted to an implementation using a future file by moving the tag array and tag logic from the register file to the future file. The description of the future file in Section 5.1.5 assumed that the future file contains tags, even though the tags were not strictly necessary for recovery: recovery requires only a bit for each register to indicate whether the register file or the future file has the most recent value. Tags are included to enforce dependencies and implement renaming.

When an instruction is decoded, it accesses tags in the future file along with the operand values. If the register has one or more pending updates, the tag identifies the update value required by the decoded instruction. Once an instruction is decoded, other instructions may overwrite this instruction's source operands without being constrained by antidependencies, because the operands are copied into the instruction window. Output dependencies are handled by preventing the writing of a result into the future file if the result does not have a tag for the most recent value. Both anti- and output dependencies are handled without stalling instruction issue.

The cost of renaming with the future file is the cost of the tag array and of the future file itself. The tag array has a write port for each instruction that is decoded at the same time (for example, four write ports for a four-instruction decoder). The tag write ports are independent of the result write ports in the future file, because tags are written immediately after decode rather than after instruction completion. Also, in addition to the read ports that supply operand tags to decoded instructions, the tag array has read ports to access tags so that they can be compared to result tags before results are written into the future file, to prevent writing an old value over a new one. The number of entries in the tag array is equivalent to the number of registers in the future file. In comparison, the number of tag entries in the reorder buffer is equivalent to the number of pending updates needed to sustain performance. In our standard processor, there are 16 tag entries in the reorder buffer; there would be 32 tag entries with a future file. Also, compared to renaming with a reorder buffer, renaming with a future file requires an additional array for the future file. The reorder buffer still is needed with a future file, though in a simpler form.

6.1.4 Enforcing Dependencies with Interlocks

If we do not remove dependencies through renaming, we must use interlocks to enforce dependencies. Recall that an interlock simply delays the execution of an instruction until the instruction is free of dependencies. There are two ways to prevent an instruction from being executed: one way is to prevent the instruction from being decoded, and the other is to prevent the instruction from being issued. In scoreboarding, we saw an example of an interlock that prevents an instruction from being decoded when the instruction has an output dependency. We could take the same approach with antidependencies and stall the instruction decoder when a decoded instruction would overwrite the operand values of one or more instructions in the window. However, there is good reason to expect that this would limit performance a bit more than scoreboarding. For example, consider the following code sequence:

```
R3 := R3 op R5      (1)
R4 := R3 + 1        (2)
R3 := R5 + 1        (3)
R7 := R3 op R4      (4)
```

If the third instruction stalls the decoder because of the antidependency on the second instruction, we would expect that the decoder would stall longer than for the output dependency on the first instruction. Instead of just waiting on the first instruction to complete, the third instruction would have to wait also on the second instruction to be issued. Two writes to the same register often have at least one intervening use of the written value, and enforcing the antidependency on the intervening use is a more severe restriction than enforcing the output dependency on the previous write.

Simulation confirms this. Stalling the decoder for antidependencies has lower performance than scoreboarding: lower by .7% with a two-instruction decoder and by 3% with a four-instruction decoder. From a performance standpoint, we are going in the wrong direction. Interlocking by stalling the decoder for a data dependency or storage conflict prevents the processor from seeing any subsequent instructions, though some of these instructions might be independent and execute while the interlock is resolved. To improve performance over scoreboarding, we must move all interlocks from the decoder to the instruction window. Torng [1984] and Acosta et al. [1986] describe an implementation of an instruction window, called a *dispatch stack*, that does just this.

The dispatch stack is an instruction window that augments each instruction in the window with dependency counts (the dispatch stack is a central window, but the same dependency techniques can apply to reservation stations). There is a dependency count associated with the source register of each instruction in the window, giving the number of pending prior updates to the source register and thus the number of updates that must be completed before all possible true dependencies are removed. There are two similar dependency counts associated with the destination register of each instruction in the window, giving both the number of pending prior uses of the register (which is the number of antidependencies) and the number of pending prior updates to the register (which is the number of output dependencies). When an instruction is decoded and loaded into the dispatch stack, the dependency counts are set by comparing the instruction's register identifiers with the register identifiers of all instructions already in the dispatch stack. These comparisons require five comparators per loaded instruction per instruction in the dispatch stack. As instructions complete, the dependency counts of instructions that are still in the window are decremented based on the source- and destination-register identifiers of completing instructions (the counts are decremented by a variable amount, depending on the number of instructions completed). Decrementing the counts requires another five comparators per completed instruction per instruction in the dispatch stack. An instruction is independent when all of its counts are zero.

The use of counts avoids having to compare all instructions in the dispatch stack to all other instructions on every cycle. It is simpler to compare decoded instructions with instructions already in the window, keeping dependency information in the window in the form of counts, than it is to constantly rederive all dependency information. The dependency counts permit the cost of the dependency hardware to be directly proportional to the number of instructions in the processor, rather than the square of the number of instructions as would apply if the dependency information were derived from scratch on every cycle. The dispatch stack still, however, requires much hardware in the form of counters and comparators: five counters per instruction in the window, five comparators per decoded instruction per instruction in the window, and five comparators per completed instruction per instruction in the window. In comparison, renaming with a reorder buffer requires no counters, no more than two comparators per decoded instruction per instruction in the window to map instruction operands (fewer comparators are required if register-port arbitration is used), and no more than three comparators per completed instruction per instruction in the window (two comparators in the reservation stations to forward results to waiting instructions and one in the reorder buffer to write the proper entry).

The dispatch stack requires even more associative logic than renaming to check dependencies, because it checks for every possible type of dependency rather than just true dependencies. This is unfortunate, because the comparators used to detect dependencies and to update dependency counts could be put to better use by implementing renaming. Figure 6-3 compares the performance of the dispatch stack, with the set of bars labeled "d-stack," in relation to the performance of renaming and scoreboarding shown previously in Figure 6-2. Eliminating the decoder stalls for output dependencies improves performance compared to scoreboarding, but renaming still has an advantage of about 6% with a two-instruction decoder and about 15% with a four-instruction decoder.

We have an obviously bad choice here. The dispatch stack does not reduce complexity relative to renaming in return for the reduction in performance relative to renaming. Both the dispatch stack and renaming track multiple updates to registers. Renaming has the advantage, though, that it is not concerned with detecting anti- or output dependencies nor with holding instruction issue until these dependencies are resolved. Renaming also does not require five counters per instruction in the window. Renaming requires additional hardware registers to be renamed, but the reorder buffer supplies this storage handily, as well as allowing restart after exceptions and mispredicted branches. It is good to trade storage in return for simpler control logic, especially since the reorder buffer storage is in a sense "free." The dispatch stack is a victim of the inherent complexity of dependency checking.

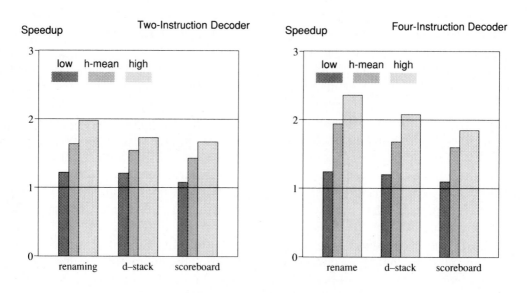

Figure 6-3. Performance of Dispatch Stack

We possibly could improve the design of the dispatch stack by reducing the number of dependencies checked. Two proposals of Tjaden and Flynn [1970, 1973] rely on the interlocks for antidependencies to also enforce output dependencies. As we have seen, two updates of a register usually are separated by an intervening use of the first updated value, as in the example at the beginning of this section. The use of the value often causes the second update to have an antidependency. Interlocking the antidependency of the second update makes it impossible for the second update to complete before the first update, because the instruction that uses the first update and creates the antidependency cannot possibly issue until the first update is complete. We could thus remove one comparator and counter from each group of five– the comparator and counter that are used to check output dependencies. However, this is a marginal improvement and relies on a dangerous assumption: that, in every possible path through the code, a register will be used at least once before it is written again. It seems likely that values are not necessarily used along every path. For example, the code at the bottom of a software loop might compute a value for the next iteration of the loop; when the loop terminates, this value is discarded and the register reused without an intervening instruction reading the old register value. It is unwise to place restrictions on this sort of code for a small hardware improvement.

The dispatch stack holds little promise, but we have not exhausted all possibilities in our quest for an alternative to renaming. In the next few sections, we will look for other ways to implement interlocks in the window or to avoid them altogether.

6.1.5 Copying Operands to Avoid Antidependencies

As we have seen in our standard processor model, in scoreboarding, and in Tomasulo's algorithm with a future file, antidependencies can be avoided altogether by copying operands to the instruction window (for example, to the reservation stations) during instruction decode. In this manner, the operands cannot be overwritten by subsequent register updates. We can copy operands to eliminate antidependencies in any approach, independent of register renaming.

An argument against copying operands is that it requires additional storage to hold copies of operands in the window. However, considerations of simplicity outweigh any concern over the extra storage for operand copies. Copying operands simplifies processor hardware because it permits a register access to be performed only once, during instruction decode. Without copying, once an instruction in the window becomes ready to issue, the processor must reaccess the instruction's operand registers before issuing the instruction, adding at least one cycle to the operation latency. The register file also must have enough ports to satisfy both instruction decoding and instruction issuing. The alternative to copying operands is to interlock antidependencies, but the comparators and/or counters required for these interlocks are costly, considering the number of combinations of source and result registers to be compared. These interlocks are also somewhat unnatural, because they are re-

---

leased by broadcasting two source-register identifiers per instruction to all other waiting instructions. There is no other purpose in broadcasting source-register identifiers. Interlocks for true and output dependencies, in contrast, are released by broadcasting a single result-register identifier per instruction as the instruction completes–this identifier must be broadcast anyway to supply operand values to waiting instructions. A more subtle benefit of copying operands is that it allows the processor to forget about how it accessed an operand once the operand has been accessed the first time. Without copying, the processor must be careful about moving the location or the mapping of an operand value. With copying, the source of the operand value is irrelevant after decode–this is why the value can be overwritten without concern for antidependencies.

Any approach involving the copying of operands must correctly handle the situation where the operand value is not available when it is accessed. We have already dismissed the possibility of stalling the decoder in this situation, because we are exploring the implications of out-of-order issue. A common solution is to supply a tag for the operand rather than the operand itself. This tag is simply a means for the hardware to identify which value the instruction requires, so that, when the operand value is produced, it can be matched to the instruction. If there can be only one pending update to a register, the register identifier can serve as a tag (as with scoreboarding). If there can be more than one pending update to a register (as with renaming), there must be a mechanism for allocating result tags and insuring uniqueness. Note that the benefits just listed of copying operands also apply when the processor copies an operand tag to the window instead of an operand value.

There is no good reason to interlock antidependencies, either in the decoder or the window, because it is too easy to copy operands. This leads us to the matter of output dependencies and one final alternative to renaming. Instead of using renaming to remove output dependencies, we can use interlocks in the window to enforce output dependencies. We are led to this approach along two different paths: revising scoreboarding to move the output-dependency interlocks from the decoder to the window or revising the dispatch stack to remove antidependency interlocks from the window. Despite this apparent obviousness, I have not been able to find an existing reference to this approach, so I have invented a name for it: *partial renaming*.

6.1.6 Partial Renaming

The dependency mechanisms described so far suggest a technique that has not been proposed in the published literature. This technique is suggested by noting that the greatest performance disadvantage of scoreboarding is due to stalling the decoder for output dependencies and that a significant portion of the hardware cost of the dispatch stack is due to interlocking antidependencies. An alternative to both approaches is to allow multiple pending updates of registers to avoid stalling the decoder for output dependencies, but to handle antidependencies by copying operands (or tags) during decode. An instruction in the window is not issued until it is free of output dependencies, so the updates to each register are performed in the same order in which they would be performed with in-order completion, except that updates for different registers are out of order with respect to each other. This alternative has almost all of the capabilities of register renaming, lacking only the capability to issue instructions so that updates to the same register occur out of order. This alternative is called partial renaming because it is close in capability to register renaming.

Figure 6-4 shows the performance of partial renaming, with the set of bars labeled "partial," in relation to the alternatives shown previously in Figure 6-3. Of all alternatives shown, partial renaming comes closest in performance to register renaming. Compared to renaming, there is a 1% performance reduction for a two-instruction decoder and a 2% performance reduction for a four-instruction decoder. The slight performance disadvantage indicates that out-of-order update of any given register is not very important. There already is a great deal of sequential ordering on the updates because of the nature of the code. It is likely that dependencies will cause updates to individual registers to occur in order anyway. It is also likely that completing an older update will release more instructions for issue than completing a newer update, and thus it is more likely that the older update will have a greater benefit to performance. Forcing individual registers to be updated in order, using interlocks, incurs little performance penalty.

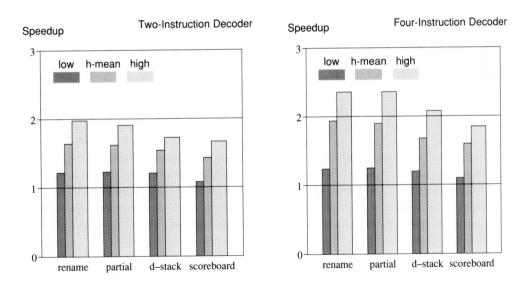

Figure 6-4. Performance Advantage of Partial Renaming

Still, we must consider the cost of partial renaming. Like renaming, partial renaming must track multiple pending updates to registers, and, to handle true dependencies correctly, requires logic to allocate and distribute tags. But partial renaming has an additional use for the tags, because these are used to sort out the various updates to the registers as well as the various uses of the register values. Partial renaming requires that, during decode, the tag storage be accessed using source-register identifiers to obtain tags for operand values *and* that the tag storage be accessed using result-register identifiers to obtain tags for the most recent update values. The latter tags are used to interlock in the window if there is a pending update to the register, enforcing output dependencies. Accessing tags thus requires a number of read ports on the tag storage equal to the number of source operands read during decode *plus* the number of results of decoded instructions. Also, interlocking instruction issue for output dependencies requires an additional comparator in each window entry to detect when a pending update is complete, freeing the instruction of its output dependency.

Partial renaming tempts us to use a future file rather than associative logic in the reorder buffer, because the reorder buffer implements renaming quite naturally and makes partial renaming rather pointless. Therefore, the cost of partial renaming also includes the cost of an additional array for the future file and the tag logic associated with this file. It is important to emphasize that we have not eliminated the tag array with partial renaming. We have, however, simplified the tag logic because the processor no longer has to check that an update to a register is the most recent update. An update, by design, is the most recent, and the future file does not have to enforce output dependencies because the window enforces output dependencies. Unfortunately, we have really only moved this complexity, because now we need comparators in the window to enforce output dependencies. We also have made the cost greater, because checking that an update is the most recent, using the tag array, takes only one comparator per update, if the tag entries of the updated registers are read from the array before the comparison. Enforcing output dependencies in the window takes one comparator per window entry. There are many fewer updates per cycle than there are instruction-window entries—for example, in our standard processor model there are two updates per cycle and 34 window entries (reservation stations). Hence we need fewer comparators to enforce output dependencies during register update than to enforce them in the instruction window.

With a four-instruction decoder, implementing partial renaming in our standard processor organization requires a 32-entry tag array having 4 write ports and 8 read ports (each entry consists of at least a 4-bit tag and 1 valid bit): 34 4-bit comparators in the reservation stations to enforce output dependencies: and a duplicate of the register-file array to implement the future file. This hardware is cumbersome, and its object is only to avoid associative lookup in the reorder buffer. Much of the reorder-buffer structure still is present. We also must consider the impact of the future file on the penalty of mispredicted branches (see Figure 5-8, page 99). Since partial renaming has lower performance and requires more hardware, it has little advantage over renaming with a reorder buffer.

So far, we have examined every apparent alternative to renaming with a reorder buffer and have found no better alternative. This result is partially due to the way we have constrained the problem. Underlying the discussion of dependencies has been the assumption that the processor performs out-of-order issue and already has a reorder buffer for recovering from mispredicted branches. Out-of-order issue makes it unacceptable to stall the decoder for dependencies. If the processor has an instruction window, it is inconsistent to limit the lookahead capability of the processor by interlocking the decoder. There are then only two alternatives: implement anti- and output-dependency interlocks in the window or remove these altogether with renaming. The inherent complexity of dependency detection, and the fact that the processor already has a reorder buffer, make renaming the most attractive alternative. Renaming does not eliminate the complexity of enforcing dependencies, but limits the complexity of the dependency logic in the window by eliminating all but true dependencies from consideration.

6.1.7 Special Registers and Instruction Side Effects

Some processor instructions do not read and write only general-purpose registers, but also use and modify values in special-purpose registers, such as status registers. This is especially characteristic of CISC instruction sets, but RISC instruction sets also have this characteristic in more limited cases. Because these uses of special state are often implicit in the instruction, rather than being specified by an explicit register identifier, renaming is very difficult to implement. And, at least in RISC processors, such uses are infrequent and do not justify the additional complexity to the renaming hardware. To enforce dependencies for such cases, it is best to consider all special-purpose state as residing in one register and interlock all accesses to this register at the decoder. This simple approach may stall the decoder more than necessary, because it cannot differentiate one special-purpose register from another. However, because such accesses are infrequent, performance is not reduced very much compared to a technique that distinguishes the registers from one another.

Some modifications to special processor state are so global in their effect, such as those that change the operating mode, that it is not worthwhile to be concerned about which instructions actually use the state. It is so likely that there is an instruction in the window whose execution will be incorrectly affected that it is best just to assume that such an instruction is indeed in the window. For these cases, it is best to use a technique called *serialization*. Serialization stalls the decoding of the special instruction until the window has been emptied of *all* instructions, these instructions are complete, and the reorder buffer is empty. Following this, the special instruction can be decoded and executed without possibly having an incorrect effect on preceding instructions. To insure that following instructions are also executed correctly, the processor can also wait until the special instruction is complete before decoding or issuing any subsequent instruction.

6.2 RESULT BUSES AND ARBITRATION

In the standard processor model, two buses are used to carry results from the integer and floating-point functional units to the respective reorder buffers. Even at high levels of performance, the utilization of two result buses, or the fraction of their capacity that is actually used, is about 50-55%. Figure 6-5 shows, for various numbers of integer result buses in the standard processor, the average bus utilization and average number of waiting results for the sample benchmarks. The principal effect of increasing the number of buses beyond two is to reduce short-term bus contention, as indicated by the reduction in the average number of results waiting for a result bus. However, as Figure 6-6 shows, eliminating this contention has a negligible benefit to performance (less that 1% going from two to three buses).

Since there are fewer result buses than functional units, the functional units must arbitrate for result buses. Many arbitration schemes can be implemented. In the standard processor model, the functional units request use of a result bus one cycle in advance of use. An arbiter decides, in any given cycle, which functional units will be allowed to gate results onto the result buses in the next cycle. There are two

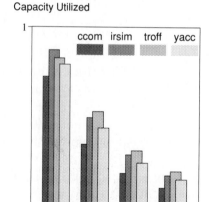
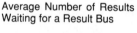
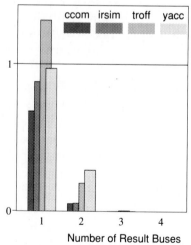

Figure 6-5. Integer Result-Bus Utilization as the Number of Buses Is Varied

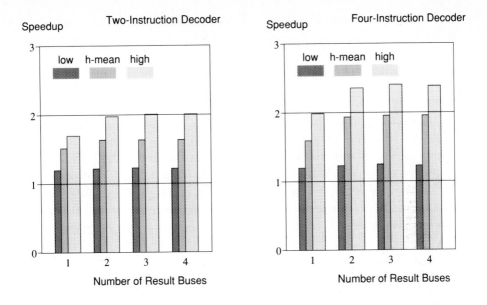

Figure 6-6. Performance Effect of the Number of Result Buses

separate arbiters–one for integer results and one for floating-point results. Priority is given to those requests that have been active for more than one cycle, then to requests that have become active in the current cycle. The integer functional units are prioritized, in decreasing order of priority, as follows: ALU, shifter, branch (for the return addresses of procedure calls), and loads. The floating-point functional units are prioritized as follows: add, multiply, divide, and convert. Prioritizing old requests over new ones helps to prevent *starvation*–that is, the possibility that a request will not be honored for many cycles–with a slight increase in arbiter complexity. The arbiter not only decides which functional units are to be granted use of the result buses, but also which bus is to be used by which functional unit.

If a functional unit requests a result bus but is not granted use, instruction issue is suspended at that functional unit until the bus is granted, though the reservation station can continue to accept instructions until it is full. Thus, a functional unit suffers start-up delay after experiencing contention for a result bus; this additional latency prevents result buses from being 100% utilized even when more than one result, on the average, is waiting for a bus (see the plot for *troff* in Figure 6-5). The advantage of suspending all instruction issue, though, is that all functional-unit pipeline stages can be clocked with a common signal. Allowing earlier pipeline stages to operate while later stages are halted complicates the clocking circuitry. Figure 6-6 indicates that reducing the effect of bus contention–by adding a third bus–yields a negligible improvement in performance. Hence, reducing the effects of bus contention by more exotic arbitration or additional pipeline buffering is unwarranted.

6.3 RESULT FORWARDING

Result forwarding supplies operands that were not available during decode directly to waiting instructions in the reservation stations. This resolves true dependencies that could not be resolved during decode. Forwarding avoids the latency caused by writing, then reading the operands from the register file or reorder buffer and avoids the additional ports that would be required to allow the reading of operands both during decode and before issue. Figure 6-7 shows, for the sample benchmarks, the distribution of the number of operands supplied by forwarding per completed result. The horizontal axis is a count of the number of reservation-station entries receiving the result as an input operand. About two-thirds of all results are forwarded to one waiting operand, and about one-sixth of all results are forwarded to more than one operand. The high proportion of forwarded results indicates that forwarding is an important means of supplying operands.

The primary cost of forwarding is the associative logic in each reservation-station entry to detect, by comparing result tags to operand tags, that a required operand is on a result bus. Our standard processor organization requires 60 four-bit comparators in the reservation stations to implement forwarding–two comparators for

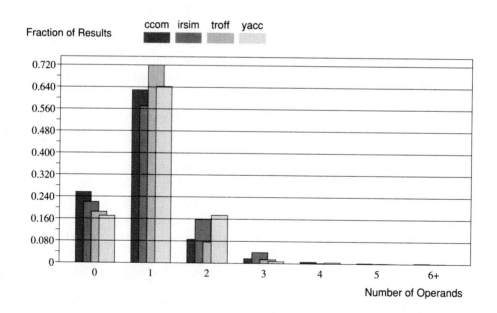

Figure 6-7. Distribution of Number of Operands Supplied by Forwarding, per Result

Superscalar Microprocessor Design

most instructions but only one for load instructions because loads have only one register-based operand. The number of bits in the tags is determined by the number of entries in the reorder buffer, because each tag must uniquely identify an entry in the reorder buffer.

The *direct tag search* algorithm proposed by Weiss and Smith [1984] eliminates the tag comparators in the reservation stations needed for matching result values to operand values. Direct tag search is used in a hardware organization very similar to our standard model, except that there is a *tag table* indexed by result tags to perform the routing function that is performed, in our model, by the reservation-station comparators (see Figure 6-8a). In direct tag search, there can be only one reference in some reservation station to a given tagged value; a second attempt to reference this tagged value stalls instruction decoding (for readers now trained to react to decoder interlocks, we will return to this stall in a moment). When an operand tag is placed into a reservation station, an identifier for the reservation station is placed into the tag table. When the result corresponding to this tag is produced, the result tag is used to access the tag table, and the reservation-station identifier from the table is used to route the result value to the reservation station.

We can examine the performance of direct tag search using an implementation that is more general than Weiss and Smith's (see Figure 6-8b). In this implementation, the tag table maintains–for each result tag–a *list* of operands that are waiting on the corresponding result. Weiss and Smith, in effect, proposed a list with a single item, but a list with several items is more general because it allows a result to be forwarded to more than one reservation station. When a decoded instruction depends on an unavailable result, the appropriate reservation-station identifier is placed on

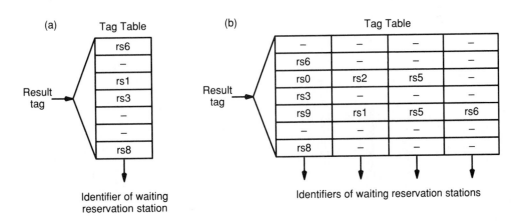

Figure 6-8. Direct Tag Search for Forwarding
(a) Weiss and Smith's Proposal (b) More General Implementation

the list of waiting operands. If the list is full, the decoder stalls–the larger the list, the less frequent we would expect decoder stalls to be. When the result becomes available, its tag is used to access the list, and the result is forwarded directly to the reservation stations on the list.

Figure 6-9 indicates the performance of direct tag search for various numbers of list entries. For best performance, at least two entries are required for each result. Unfortunately, any implementation with more than one entry encounters difficulty allocating entries and detecting that the list is full. Furthermore, even an implementation with one list entry requires a storage array for the tag table. In the standard processor, this table would have 16 entries of 6 bits each, with 8 ports for writing reservation-station identifiers and 2 ports for reading reservation-station identifiers. There also would be 60 decoding circuits to gate the results into the proper reservation-station entries. Direct tag search was proposed for a scalar processor and does not save hardware in the superscalar processor because of the larger number of instructions decoded and completed per cycle.

In Figure 6-9, the performance with zero list entries is the performance that would result if forwarding were not implemented. The decoder stalls for any true dependency in this case, because, without forwarding, there is no way for an instruction to obtain an operand after decode if the operand is not available during decode. Thus, we might expect that this alternative would have about the same performance

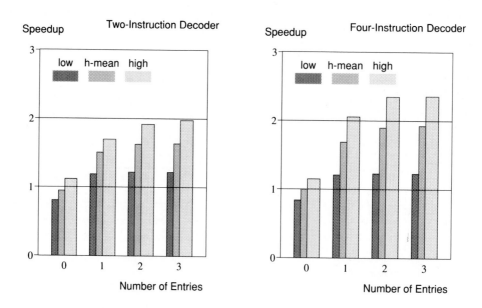

Figure 6-9. Performance of Direct Tag Search for Various Numbers of List Entries

as in-order issue, or an average speedup of about 1.2. However, without forwarding, there is no advantage at all to the superscalar processor; average performance is slightly lower than or about the same as the scalar processor, depending on the width of the decoder. With forwarding, the processor can issue stalled instructions a cycle earlier than without forwarding, because the processor does not have to wait for an operand to be written into, the read from the reorder buffer or register file. The lower performance without forwarding shows that the machine parallelism of the superscalar processor does not compensate for the data-communication delay caused by a lack of forwarding. This further illustrates the value of forwarding and emphasizes that exploiting instruction parallelism requires efficient communication of data among different computations.

Forwarding not only takes a large amount of hardware, but also creates long logic-delay paths. Once a result is valid on a result bus, it is necessary to detect that this is a required result and gate the result into the reservation station, so that the functional unit can begin operation while the result is written into the reorder buffer. To avoid having the forwarding delay appear as part of the functional-unit delay, and to allow data paths to be set up in time, the actual tag comparison should be performed in the cycle preceding the cycle in which the result is driven. This in turn implies that the tags for the two results should appear on the tag buses a cycle before the results appear on the result buses. Providing result tags early is difficult, because it requires that result-bus arbitration be performed very early in the cycle.

6.4 SUPPLYING INSTRUCTION OPERANDS: OBSERVATIONS

If software alone cannot enforce dependencies, renaming with a reorder buffer is the best hardware alternative for resolving operand dependencies in a superscalar processor. Unfortunately, dependency analysis is fundamentally hard, especially when multiple instructions are involved. All the alternatives to renaming explored in this chapter require an equivalent or greater amount of hardware in comparison to renaming, with reduced performance. The buses, associative logic, and multiple-port arrays required by register dataflow are unattractive, but are essential to the design. Without the associative tag logic for forwarding, the superscalar processor has worse performance than does a scalar processor. Adding a future file to the architecture can eliminate the associative lookup in the reorder buffer, but a future file also requires tag comparisons in the writeback logic to implement renaming and incurs the cost of the future file itself. The tag comparisons in the writeback logic can be eliminated by partial renaming, but the logic to guarantee that registers are updated in order is more complex than the eliminated logic. Scoreboarding, because it does not implement renaming, can be used to eliminate the associative lookup in the reorder buffer *and* the tag comparisons in the writeback logic. But scoreboarding still requires the scoreboard logic itself, and scoreboarding yields the poorest performance of any alternative explored in this chapter. Also, scoreboarding requires a future file in addition to a reorder buffer to recover from mispredicted branches, because a future file is required to supply the architectural state whenever there is no associative lookup in the reorder buffer. Considering the cost of the future file, any approach using a future file has questionable value.

The cost of the dependency and forwarding hardware should be viewed as a more or less fixed cost of attaining high performance. Of course, this can lead to the conclusion that such performance is too ambitious. The argument offered here is only that there are no simple hardware alternatives to renaming yet discovered that yield correct operation with good performance. Chapter 7 considers other ways to reduce the amount of hardware by focusing on alternatives to reservation stations.

Chapter 7

Out-of-Order Issue

The peak instruction-execution rate of a superscalar processor can be almost twice the average instruction-execution rate, as shown in Figure 3-8 (page 47). This is in contrast to a scalar RISC processor, which typically has an average execution rate that is within 30% of the peak rate. A simpleminded approach to superscalar design provides enough hardware to support the peak instruction-execution rate. However, this hardware is expensive, and the difference between the average and peak execution rates implies that the hardware is underutilized.

A more cost-effective approach to processor design is to constrain the hardware carefully in light of the average instruction-execution rate. We have seen several examples of this approach in preceding chapters. In Chapter 4, we demonstrated that the number of register-file read ports to satisfy a four-instruction decoder can be constrained to four ports–at first blush, only enough ports to sustain a two-instruction decoder. Similarly, in Chapter 6 we demonstrated that the tag logic associated with the future file in Tomasulo's algorithm can be consolidated into a reorder buffer that requires less tag hardware because it tracks only pending register updates rather than all possible updates to all registers.

In this chapter, we turn to the underutilization of the reservation stations in our standard processor model (see Table 3-3, page 48). These reservation stations–used to obtain all results so far–have a total capacity of 34 instructions. However, Chapter 5 showed that the reorder buffer should have 12 to 16 locations. Since the size of the reorder buffer is determined by instructions in the window *and* in the pipeline stages of the functional units, this indicates that the instruction window contains fewer than about 16 instructions on the average. There seem to be many more reservation-station entries than are really required. Reservation stations were chosen for the instruction window without much justification in Section 3.4.2. Perhaps this is a source of hardware savings.

The problem of underutilization of the reservation stations is overcome by consolidating them into a smaller *central window*. The central window holds all unissued instructions, regardless of the functional units that execute the instructions, as illustrated in Figure 7-1 (only the integer functional units are shown in Figure 7-1; the central window may be duplicated in the floating-point unit, depending on the implementation). For a given level of performance, the central window uses much

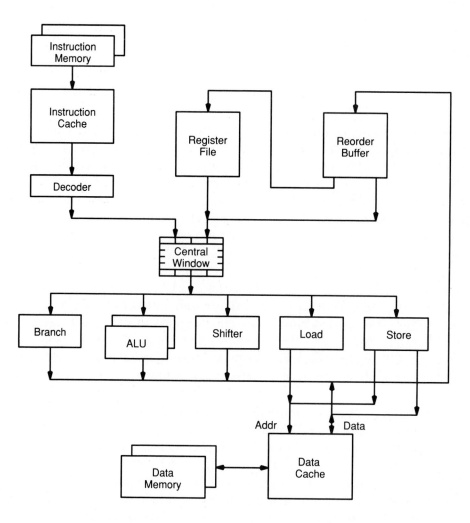

Figure 7-1. Location of Central Window in Processor (Integer Unit)

less storage than reservation stations, less control hardware, and less logic for resolving dependencies and forwarding results.

This chapter presents the motivation for the reservation-station sizes given in Table 3-1, explores a simplification to the reservation stations, and then focuses on the implementation of a central window to save hardware. We will examine two published central-window proposals: the *dispatch stack* proposed by Torng [1984] and the *register update unit* of Sohi and Vajapeyam [1987]. The dispatch stack and register update unit are complete proposals for out-of-order issue and thus include dependency mechanisms. In this chapter, we are not interested in the dependency mechanisms, but rather in the algorithms used to place instructions into the window, to issue instructions from the window, and to remove instructions from the window. Both the dispatch stack and the register update unit have disadvantages: the dispatch stack is complex, and the register update unit has relatively poor performance. In both proposals, the disadvantages are the result of using the central window to serve purposes that are better served by a reorder buffer. A third proposal presented in this chapter relies on the existence of a reorder buffer and its dependency hardware to simplify the central window.

7.1 RESERVATION STATIONS

Reservation stations partition the instruction window by functional unit, as shown in Figure 3-7 (page 45). This partition helps to simplify the control logic at each reservation station but causes the total number of entries in the reservation stations to be much larger than the total number of entries in the central window, for an equivalent level of performance.

7.1.1 Reservation Station Operation

A reservation-station entry holds a decoded instruction until it is free of dependencies and the associated functional unit is free to execute the instruction. The following steps are taken to issue instructions from a reservation station:

1. Identify entries containing instructions ready for issue. An instruction is ready when it is the result of a valid decode cycle and all of its operands are valid. Operands may have been valid during decode or may become valid when results are computed and forwarded to the reservation station.

2. If more than one instruction is ready for issue, select an instruction for issue among the ready instructions.

3. Issue the selected instruction to the functional unit.

4. Deallocate the reservation-station entry containing the issued instruction so that this entry may receive a new instruction. For best utilization of the reserva-

tion station, the entry should be able to receive a new instruction from the decoder in the next cycle.

These functions are performed in any instruction window that implements out-of-order issue. Reservation stations have some advantage, though, in that these functions are partitioned and distributed in a way that simplifies the control logic. Most reservation stations perform these functions on a small number of instructions. Only the load and store reservation stations are relatively large (eight entries each), but these reservation stations are constrained to issue instructions in order at the data-cache interface, and so there is only one instruction that can be issued: the oldest instruction. The issue logic simply waits until this instruction is ready rather than selecting among all ready instructions. A reservation station also has simple deallocation logic, because it can free at most one instruction per cycle. Of course, reservation stations duplicate the control hardware at each functional unit, and this duplication is a source of hardware cost.

7.1.2 Performance Effect of Reservation-Station Size

Instruction decoding stalls if a decoded instruction requires a reservation station that is full. Reservation stations become full if either data dependencies or functional-unit conflicts cannot be resolved at a sufficient rate compared to the rate of instruction fetching and decoding. A given reservation station must be large enough to handle the majority of short-term situations where the instruction-decode rate exceeds the instruction-execution rate for the corresponding functional unit. Otherwise, the mismatch in rates stalls the decoder and prevents the decoding of instructions for other functional units, reducing the instruction bandwidth and the lookahead capability of the processor.

Figure 7-2 shows the effect of reservation-station size on the performance of the superscalar processor. To simplify the presentation of data, these results were measured with all reservation stations having the same size. With a two-instruction decoder, performance is not reduced significantly until the reservation stations are reduced to two entries each. Instruction-fetch limitations prevent a processor with a two-instruction decoder from taking advantage of more than two entries. With a four-instruction decoder, performance is noticeably reduced when reservation stations are reduced to four entries each. In reality, however, not all reservation stations require more than four entries. The decrease in performance going from eight to four entries is due primarily to the reservation station for stores, as we will see. Experimentation indicates that the reservation-stations sizes given in Table 3-3 provide good performance.

Figure 7-3 illustrates, for the sample benchmarks, the role that the reservation stations serve in providing adequate instruction bandwidth and therefore adequate lookahead capability. Figure 7-3 shows, by integer functional unit, the average instruction bandwidth lost at the decoder because of full reservation stations, as the

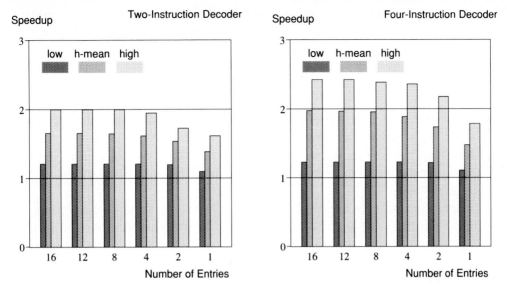

Figure 7-2. *Performance Effect of Reservation-Station Size*

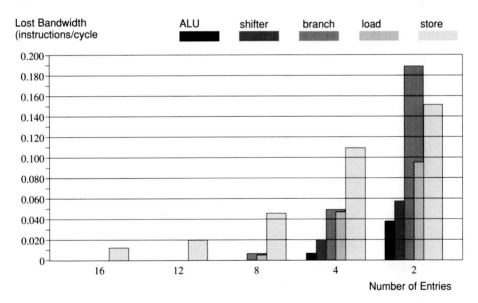

Figure 7-3. *Relative Contribution of Reservation Stations to Lost Instruction Bandwidth*

sizes of the reservation stations are reduced (the floating-point reservation stations cause negligible penalties for these benchmarks). Reservation-station size is most critical for the branch, load, and store functional units. These functional units are constrained, for correctness, to execute their instructions in the original order with respect to other instructions of the same type, and thus have the highest probability of a mismatch between instruction-decode and instruction-execution rates. The store functional unit is generally the worst in this respect. To preserve the in-order state in the external memory, stores are constrained to issue only after all previous instructions have completed, then contend with loads (with low priority) for use of the address bus to the data cache. However, a small branch reservation station causes the most severe instruction-bandwidth limitation, because branches are more frequent than stores and because the branch delay is typically three or four cycles. The processor must allow about four unexecuted branches for best performance, as we saw in Chapter 4.

Reservation stations are underutilized because they serve two purposes. The first purpose is that they implement the instruction window. As we have seen from our examination of the reorder buffer, the window requires buffering for fewer than 16 instructions. The second purpose is that the reservation stations prevent short-term demands on the respective functional units from stalling the decoder. Preventing decoder stalls takes 2 to 8 entries at each functional unit, for a total of 34 entries. The latter consideration is most important for the branch, load, and store functional units, but causes the total number of reservation-station entries for all functional units to be larger than the number required solely for the instruction window.

7.1.3 A Simpler Implementation of Reservation Stations

Before examining ways to consolidate the reservation stations into a central window, we should note that all reservation stations can operate as do the load, store, and branch reservations stations, without much loss in performance. That is, each reservation station can issue instructions in the order that they are received from the decoder. Instructions are issued in order at each functional unit, but the functional units are still out of order with respect to each other, and thus the processor still performs out-of-order issue. The advantage of this approach is that each reservation station need only examine one instruction to issue–the oldest one–rather than all instructions it currently holds. The instruction-issue logic is simpler than if the reservation station examines all instructions, because this logic need not resolve contention for the functional unit. Also, since there is only one entry that can be deallocated, the reservation station can be managed as a first-in, first-out buffer.

Figure 7-4 repeats the presentation of Figure 7-2, except that instructions are issued in order at each reservation station rather than out of order. For comparison, the performance of out-of-order issue at the reservation stations is shown as an increment to performance in each case. The performance advantage of out-of-order issue at the reservation stations is negligible in some cases. Across the entire range of reservation-station sizes, the biggest difference in average performance between the two approaches is .6% for a two-instruction decoder and 2% for a four-instruction

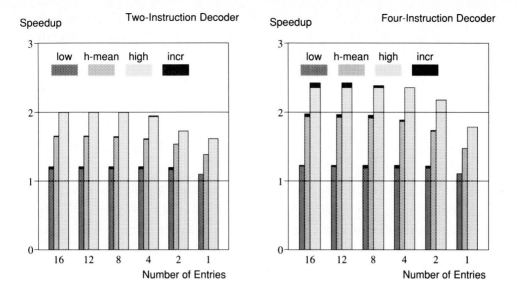

Figure 7-4. Performance Effect of Issuing Instructions In Order at Reservation Stations

decoder. The degradation is small because there is some sequential ordering on the code–we saw a similar example of this when we examined partial renaming in the previous chapter. When a dependency stalls instruction issue at a particular functional unit, it is more important for the processor to issue instructions at other functional units than to issue a newer instruction at the stalled functional unit.

There is another way to justify the fact that the performance difference is so small between in-order issue and out-of-order issue at the reservation stations. We have already constrained the load, store, and branch units to issue instructions in order. This does not leave very much else to benefit from out-of-order issue. Only the ALU instructions are frequent enough that they might benefit–except that there are two ALUs, and instructions are distributed between these so that the two oldest ALU instructions are in different reservation stations. Given that the two oldest ALU instructions can be issued out of order, it is easy to understand how there can be little benefit to out-of-order issue at the reservation stations.

7.2 IMPLEMENTING A CENTRAL INSTRUCTION WINDOW

In terms of the number of entries in the instruction window, a central window is more efficient than reservation stations, because the central window holds all instructions for issue regardless of which functional units execute the instructions. The window resolves the short-term demands on any functional unit. But the central window consolidates all instruction-issue logic into a single structure, and the logic associated with a central window is more complex than the corresponding logic in a reservation station, for several reasons:

- The central window selects among a larger number of instructions than does a reservation station. The central window examines all instructions in the window rather than a subset of these instructions as is the case with a reservation station. We just saw in the previous section that a reservation station can examine as little as a single instruction at a time while retaining most of the advantages of out-of-order issue.

- The central window must consider functional-unit requirements in selecting among ready instructions and must arbitrate among ready instructions.

- The central window can free more than one entry per cycle, so allocation and deallocation are more complex than with reservation stations.

- An entry in the central window must be able to hold instructions of any type, whereas an entry in a reservation station can be tailored to a specific set of instructions (that is, the instructions executed by the corresponding functional unit). For example, in the R2000 architecture, each window entry must provide storage for three 32-bit words, because branch instructions need three input operands: two operands are compared and the third operand specifies the branch-target address. With reservation stations, the capability to hold these three operands is required only in the branch reservation station.

This section explores two implementations of a central window–the dispatch stack [Torng 1984] and the register update unit [Sohi and Vajapeyam 1987]–and proposes a third implementation that has better performance than the register update unit without the complexity of the dispatch stack. The proposed implementation presumes the existence of a reorder buffer, whereas the dispatch stack and register update unit incorporate some of the functions of the reorder buffer and do not make this presumption. However, the reorder buffer is an important component for supporting branch prediction and register renaming: the fact that the reorder buffer can simplify the central window is a further advantage of the reorder buffer.

It is also shown in this section that the number of buses required to distribute operands from a central window is comparable to the number of buses required to distribute operands to reservation stations–negating the concerns expressed in Section 3.4.2. The central window adjusts instruction issue to account for the operand-bus limitation. This is another example of something we discovered when we reduced the number of register-file ports in Section 4.4: the buffering provided by the instruction window allows hardware to be constrained in some ways without much reduction in performance. However, arbitrating for operand buses further complicates the central window's instruction-issue logic, much as arbitrating for register-file ports complicates instruction decode.

7.2.1 The Dispatch Stack

On every cycle, the dispatch stack [Torng 1984, Acosta et al. 1986, Dwyer and Torng 1987] performs operations that are very similar to those performed in a reservation station:

1. Identify entries containing instructions ready for issue.

2. Select instructions for issue among the ready instructions, based on functional-unit requirements. This involves prioritizing among instructions that require the same functional unit.

3. Issue the selected instructions to the appropriate functional units.

4. Deallocate the dispatch-stack entries containing the issued instructions, so that these entries may receive new instructions.

Conceptually, instructions are placed at the top of the dispatch stack by the decoder and are issued from the bottom of the stack. The dispatch stack keeps unissued instructions in the order that instructions were originally decoded, with older instructions toward the bottom of the stack. This is necessary for correct operation of the dependency logic, because a given instruction can release dependencies only of instructions that follow it, not instructions that precede it. Keeping instructions in order also makes it simpler to prioritize among ready instructions, because it is easy to locate the older instructions that should be preferred over newer instructions: older instructions are located nearer the bottom of the stack. To keep instructions in order, the dispatch stack cannot just allocate entries by placing decoded instructions into freed locations: it must *compress* the window by filling freed locations with older, waiting instructions and then allocating entries for new instructions at the top of the stack, as shown in Figure 7-5. An instruction or sequence of instructions can remain in the dispatch stack for an arbitrary amount of time; other instructions flow around the waiting instructions.

Figure 7-6 shows the effect of dispatch-stack size on the performance of the superscalar processor. This performance was measured with a dispatch stack that can accept a portion of the decoded instructions when the stack does not have enough entries for all decoded instructions. Both a two-instruction and a four-instruction decoder require only an eight-entry dispatch stack for best performance. This is consistent with the size expected from the size of the reorder buffer. Thus, the dispatch stack uses about one-fourth of the storage and forwarding logic of the reservation stations. Furthermore, the instruction-issue logic is not duplicated, unless it is duplicated in separate integer and floating-point units. With the reservation stations in Table 3-3, the instruction-issue logic is duplicated 10 times.

Despite this advantage, issuing instructions with a dispatch stack is complex, because the dispatch stack keeps instructions in order. Compressing and allocating the window can occur only after the window has identified the number and location

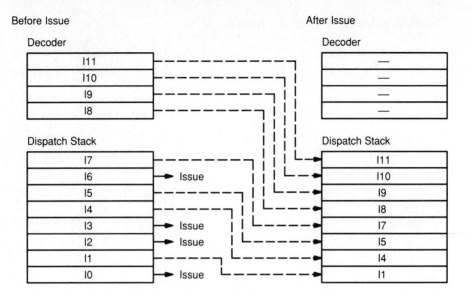

Figure 7-5. Compressing the Dispatch Stack

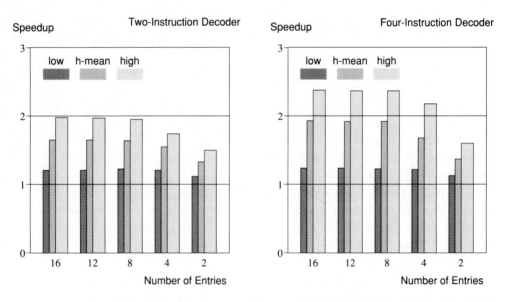

Figure 7-6. Performance Effect of Dispatch-Stack Size

of the instructions for issue. Compressing the window also requires that any entry be able to receive the contents of any newer entry and that any entry be able to receive an instruction from any decoder position. Thus, for example, an entry in an 8-location window with a 4-instruction decoder requires a maximum of 11 possible input sources, a minimum of 4, and an average of about 8. Dwyer and Torng [1987] estimate that the issue, compression, and allocation logic for a proposed implementation of an 8-entry dispatch stack consumes 30,000 gates and 150,000 transistors. This is roughly the number of transistors in many first-generation RISC processors (that is, RISC processors without on-chip caches)–a fact that argues against the dispatch stack.

7.2.2 The Register Update Unit

Although the dispatch stack requires less storage for instructions and operands than do reservation stations, the complexity of the issue, compression, and allocation hardware in a dispatch stack argues against its use. The *register update unit* [Sohi and Vajapeyam 1987] is a simpler implementation of a central window that avoids the complexity of compressing the window.

Sohi and Vajapeyam examined several extensions to Tomasulo's algorithm that provide precise interrupts. One of their proposals followed the observation that each active entry in the tag array of Tomasulo's algorithm is associated with a single instruction in a reservation station (recall that the tag array is associated with the register file to prevent old updates from overwriting newer ones). Because of this one-to-one correspondence between tag entries and reservation-station entries, the tag array and reservation stations can be combined into a single *reservation station/tag unit (RSTU)* that serves as a central window for all functional units. In a further refinement, the storage in the RSTU can be used to hold results after instruction completion and can return the results to the register file in sequential order. Consequently, the RSTU can act much like a reorder buffer as well as an instruction window. This refined configuration, called a *register update unit (RUU)*, provides precise interrupts. The RUU is operated as a FIFO buffer, again very much like the reorder buffer, with decoded instructions being placed at the top of the FIFO and result values being written to the register file from the bottom of the FIFO.

The operations performed to issue instructions in the register update unit are identical to those in the dispatch stack (except that the dependency mechanisms are different, a topic we are not concerned with here). But there is an important difference between the two windows. The register update unit allocates and frees window entries strictly in a FIFO order, as shown in Figure 7-7. Instructions are entered at the top of the FIFO, and window entries are kept in sequential order. An entry is not removed when the associated instruction is issued. Rather, an entry is removed when it reaches the bottom of the FIFO. When an entry is removed, its result (if applicable) is written to the register file. Though this method was originally proposed to aid the implementation of precise interrupts, it also makes window allocation independent of instruction issuing, using a simple algorithm.

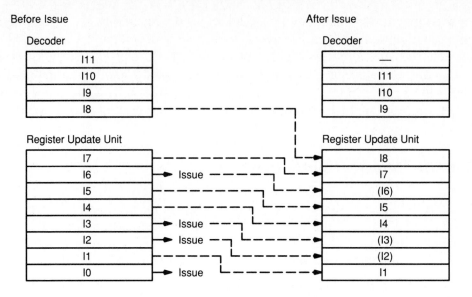

Figure 7-7. Register Update Unit Managed as a FIFO

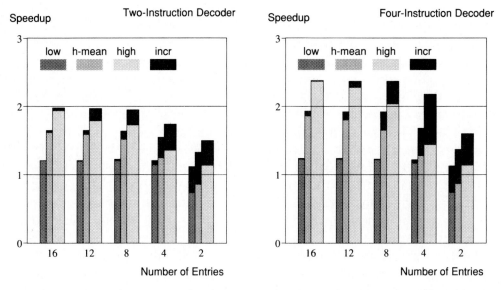

Figure 7-8. Performance Degradation of Register Update Unit Compared to Dispatch Stack

Figure 7-8 shows the performance of the register update unit, as a function of its size. The lightly shaded bars in Figure 7-8 show the speedup obtained with the register update unit, and the darkened areas show the incremental performance advantage of the dispatch stack (from Figure 7-6) for comparison. For all sizes shown, the performance of the register update unit is significantly lower than the performance of a dispatch stack of equal size. The performance disadvantage is most pronounced for smaller window sizes. Performance can be lower than that of the scalar processor, because instructions remain in the register update unit for at least one cycle before their results are written to the register file, possibly stalling the decoder for longer periods than if results are written directly to the register file (we saw a similar effect with a small reorder buffer in Figure 5-9, page 101). These results agree with results reported by Sohi and Vajapeyam: they find that a 50-entry register update unit provides the best performance, in a processor that has longer functional-unit latencies than our standard processor.

There are two reasons that the register update unit has this performance disadvantage (Figure 7-7 helps to illustrate these points):

- Window entries are used for instructions which have been issued.

- Instructions cannot remain in the window for an arbitrary amount of time without reducing performance. Rather, the decoder stalls if an instruction has not been issued when it reaches the bottom of the register update unit.

Figure 7-9 shows, for the dispatch stack described in Section 7.2.1, the average distribution of the number of cycles from the time an instruction is decoded to the time it is issued. With the register update unit, instructions at the high end of this distribution stall the decoder, because they cannot be issued and completed by the time they arrive at the bottom of the window. As we discussed in Section 7.1.2, stores experience the worst decode-to-issue delay, because they can be issued in order only after all previous instructions are complete, and then contend with loads for the data cache.

It should be emphasized that Sohi and Vajapeyam proposed the register update unit as a mechanism for implementing precise interrupts and restarting after mispredicted branches. The disadvantages cited here are due to the different objective of reducing the size of the instruction window.

7.2.3 Using a Reorder Buffer to Simplify the Central Window

Both the complexity of the dispatch stack and the relatively low performance of the register update unit can be traced to a common cause: both proposals attempt to maintain instruction order in the central window. But a reorder buffer is much better suited to maintaining the instruction order. The reorder buffer keeps dependency information in the correct order, eliminating the need to maintain this order in the window to enforce dependencies. The reorder buffer is not compressed as is the dispatch stack, and the deallocation of reorder-buffer entries is not in the critical path of instruction issue. Rather, reorder-buffer entries are allocated at the top of a FIFO and are deallocated as results are written to the register file. However, in contrast to the

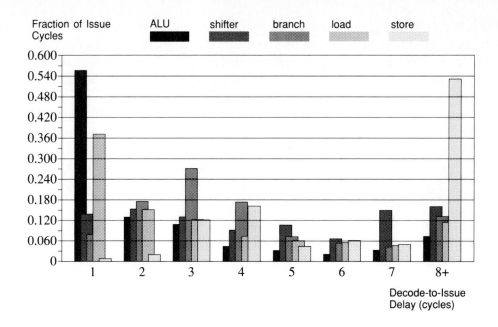

Figure 7-9. Distribution of Decode-to-Issue Delay for Various Functional Units

register update unit, which also operates as a FIFO, instructions do not stall the instruction decoder if they are not issued by the time the corresponding entry reaches the bottom of the reorder buffer. As described previously, for example, stores are simply released for issue during the writeback operation at the bottom of the reorder buffer. Several stores can be released in a single cycle, so the reorder buffer does not suffer the single-issue constraints of a store in the register update unit. Consequently, the reorder buffer does not stall instruction decoding as readily as the register update unit. The method of handling stores is the primary disadvantage of the register update unit; the primary advantage of the reorder buffer in this case is that it does not serve the dual purpose of holding instructions for issue.

In a processor with a reorder buffer, then, there is no real need to keep instructions ordered in the central window. The window need not be compressed as instructions are issued. Instead, new instructions from the decoder are simply placed into the freed locations, as Figure 7-10 shows. The principal disadvantage of this approach is that the instruction-issue logic does not know the original instruction order, and so the issue logic cannot prioritize instructions for issue based on this order. In theory, when two instructions are ready for issue at the same time and require the same functional unit, it is better to issue the older instruction first, because the older instruction is more likely to free other instructions for issue. Still, the data in Figure 7-11 suggest that this is not a very important consideration. Figure 7-11

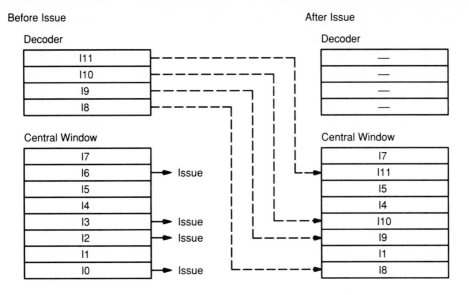

Figure 7-10. Allocating Window Locations Without Compressing

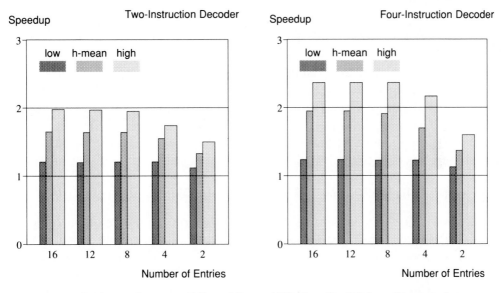

Figure 7-11. Performance Effect of Central-Window Size Without Compression

shows performance as a function of central-window size when the window is allocated without compression as shown in Figure 7-10. In this configuration, instructions are prioritized for issue based on their position in the central window, but this

position has nothing to do with the original program order. Compared to the performance of the dispatch stack, the reduction in performance caused by this less optimal prioritization is under 1% for any benchmark–the reduction is less than the measurement precision for most benchmarks. Obviously, the dependencies between instructions are sufficient to prioritize instruction issue. Issue conflicts between older and newer instructions do not occur very often. We might have already expected this from the results in Section 7.1.3, which showed that reservation stations do not reduce performance very much if they always wait for the oldest instructions to be ready and do not consider newer instructions at all.

On the other hand, we cannot unconditionally say that the ordering of instructions in the central window is unimportant. We may be measuring, as least in part, an artifact of the code as generated by a scalar compiler. As we will see in Chapters 10 and 11, there is much a compiler can do to reorder code so that the superscalar processor executes it more efficiently. The code we are measuring has not been reordered for the superscalar processor, and it could be that this is the reason we are finding that instruction order is unimportant in the central window. To the contrary, issuing instructions from the window so that older instructions are given priority might be quite helpful in gaining the full benefit from software reordering. Without access to reordered code, we cannot make a conclusive measurement.

Even though instructions may not have to be ordered in the window for performance, some instructions (loads, stores, and branches) must be issued in original program order. The central window proposed here does not preserve instruction ordering, and thus complicates issuing instructions in order. This difficulty is overcome by allocating sequencing tags to these instructions during decode if there are other, unissued instructions of the same type in the window. When a sequenced instruction is issued, it transmits its tag to all window entries, and the window entry with a matching sequencing tag is released for issue. Because there are so few window entries, this does not negate the hardware savings over the reservation stations.

7.2.4 Operand Buses from a Central Window

A concern with a central window is the potentially large amount of global (chipwide) interconnection required to supply operands to the functional units from the central window (this concern was expressed in Section 3.4.2). Distributing instructions and operands from a central window is significantly different from distributing them from reservations stations. Instructions and operands are placed into the central window using local interconnections from the decoder, reorder buffer, and register file. The window, in turn, issues these instructions and operands to the functional units using global interconnections. In contrast, instructions and operands are placed into reservation stations using global interconnections. The reservation stations, in turn, issue these instructions and operands to the functional units using local interconnections.

Section 4.4 established that a limit of four register-based operands is sufficient to supply reservation stations and does not significantly reduce performance. This allows us to approximate the number of interconnection buses required for the distribution of instructions and operands to reservation stations. However, this number of operands is sufficient because empty decoder positions and dependencies between decoded instructions reduce the demand on register-based operands during decode. If instructions and operands are issued to the functional units from a central window, these empty positions and dependencies do not exist, because the instructions are being issued for execution.

In semiconductor technology, interconnection networks are much less dense than storage elements, so the interconnection required by a central window could consume more chip area than the additional storage of the reservation stations. Because of this consideration, Figure 7-12 shows how the performance of a central window depends on the number of operand buses. A large, 16-entry window is used in order to focus on the effect of limited interconnection rather than on the effect of a limited window size. As shown, the number of operand buses can be limited to four (comparable to the number of buses adequate to distribute operands to reservation stations) with negligible effect on performance.

The reason that limiting the number of buses has so little effect is that the instruction window is able to adjust instruction issue around this limitation.

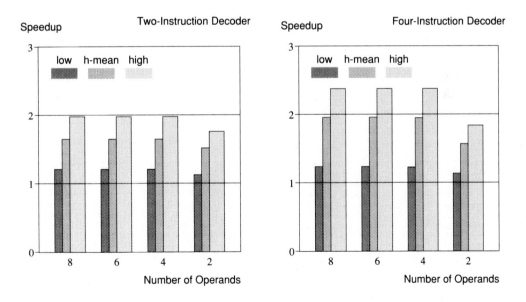

Figure 7-12. *Performance Effect of Limiting Operand Buses from a 16-Entry Central Window*

Figure 7-13 illustrates this point by comparing the instruction-issue distribution from a central window having eight operand buses to the same distribution from a window having four operand buses. As shown, limiting the number of buses reduces the percentage of cycles in which a large number of instructions are issued, but the instructions not issued are likely to be issued later with other instructions, and the average execution rate is maintained.

7.2.5 The Complexity of a Central Window

The difficulty of operand distribution from a central window is not that too many buses are required, but the fact that the instruction-issue logic must arbitrate for operand buses and conditionally issue instructions based on this arbitration. The arbitration for operand buses is conceptually similar to the arbitration for register ports discussed in Section 4.4, except that there are as many as 16 operands contending for the 4 shared buses with an 8-entry window, in contrast to 8 operands contending for 4 shared ports with a four-instruction decoder. Hence, arbitration in the central window is much more complex than arbitration in the decoder. Furthermore, in the window, it is more difficult to determine which of the contenders are actually able to use the buses. For example, to determine which instructions are allowed to use operand buses, it is first necessary to determine which instructions are ready to be issued and which functional units are ready to accept new instructions. Following this, func-

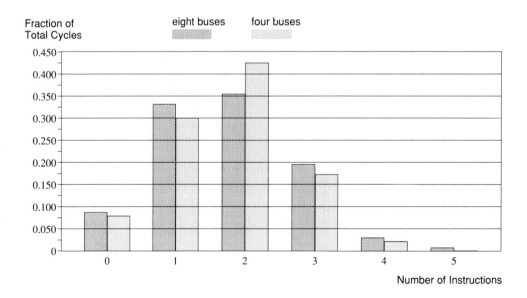

Figure 7-13. Change in Instruction-Issue Distribution from a
16-Entry Central Window, as Operand Buses Are Limited

tional-unit conflicts among ready instructions must be resolved before instructions arbitrate for operand buses. Fortunately, there is somewhat more time to accomplish these operations than there is with register-port arbitration during decode, because register-point arbitration typically must be completed halfway through a processor cycle.

The longest identifiable path in the instruction-issue logic arises when an instruction is about to be made ready by a result that will be forwarded. In this case, the result tag is valid the cycle before the result is valid, and a tag comparison in the instruction window readies the instruction for issue. Once the issue logic determines that the instruction is ready, the instruction participates in functional-unit arbitration, then operand-bus arbitration. If the instruction can be issued, the location it occupies is also made available for allocation to a decoded instruction. Table 7-1 gives an estimate of the number of stages required for each of these operations, using serial arbitration as described in Section 4.4 for the complex arbiters and assuming no special circuit structures (such as wired logic). The 16 stages required are within the capabilities of most technologies.

It is interesting to note that four of the operations in Table 7-1 are concerned with resource arbitration and allocation. That these operations have also been examined within the contexts of instruction decoding (register-port allocation) and dependency checking (tag and reorder-buffer allocation) suggests that hardware structures for arbitration and allocation are important areas for detailed investigation. Sustaining an instruction throughput of more than one instruction per cycle with a reasonably small amount of hardware creates many situations where multiple operations are contending for multiple, shared resources. The speed and complexity of the hardware for resolving resource contention is easily as important as any other implementation consideration we have examined or will examine, but it is impossible to examine this hardware without knowing precisely the details of the implementation.

Table 7-1. Critical Path for Central-Window Issuing

Function in Path	Number of Stages
result-bus arbitration	2
drive tag buses	1
compare result tags to operand tags and determine ready instructions	3
functional-unit arbitration	2
operand-bus arbitration	4
allocate free window entries to decoded instructions	4
Total	16

7.3   OUT-OF-ORDER ISSUE: OBSERVATIONS

Out-of-order issue can be accomplished either by reservation stations or by a central instruction window.   Reservation stations hold instructions for issue near to the corresponding functional unit, distributing and partitioning the tasks involved in issuing instructions.   In contrast, the central window maintains all pending instructions in one central location, examining all instructions in every cycle to select candidates for issue. The central window is more storage efficient and consolidates all the issue logic into a single unit. However, the issue logic is more complex than the issue logic of any given reservation station, primarily because there are more instructions in the central window than in a reservation station and because functional-unit conflicts must be resolved by the central window.

The implementation of a central window is simpler if we rely on a reorder buffer to maintain the instruction order. This is one of several examples we have seen of how the proper decomposition of the superscalar design reduces hardware and increases performance. There is much synergy between hardware components in the superscalar processor. For example, the instruction window aids branch prediction by providing storage for predicted instructions. The reorder buffer aids branch prediction by providing a recovery mechanism, aids out-of-order issue by implementing renaming, and, now, aids the implementation of a central window by eliminating the need to keep instructions ordered in the window.

Despite this, we cannot say that the hardware proposed so far is in any way simple. At this point, we can only hope that the complexity is worthwhile. We will examine one final hardware topic in the next chapter, then turn to the topic of complexity and explore other, possibly more attractive, alternatives.

Chapter 8

Memory Dataflow

In many RISC processors, including the R2000, loads and stores are the only operations, other than floating-point operations, that have result latencies greater than one cycle. Though the latencies of loads and stores are not as great as the latencies of floating-point operations (unless a cache miss occurs), the latencies of loads and stores affect performance more than the latencies of floating-point operations because loads and stores are more frequent than floating-point operations.

This chapter surveys various techniques for performing loads and stores in a superscalar processor. It examines alternative mechanisms for ordering loads and stores, computing and translating addresses, and checking memory dependencies. Furthermore, this chapter considers how these mechanisms affect performance and considers the hardware implications.

Reservation stations provide much leeway in the implementation of loads and stores. For this reason, we begin this chapter by considering load and store techniques using reservation stations, to illustrate general principles. However, reservation stations do not yield the most efficient implementation of loads and stores. Hence, we also examine in this chapter an implementation of loads and stores using a central instruction window—or at least using a common reservation station for both loads and stores. This simplifies the hardware with a slight reduction in performance.

Throughout this chapter, we assume that the timing and pipelining of address operations (computation and translation) and of the data-cache interface are comparable to those of the R2000 processor. This avoids distorting the results with effects that are not related to the superscalar processor but is not meant to imply that these are the only approaches to implementing these functions. For example, the overhead involved in issuing a load or store in the superscalar processor may prevent the computation and translation of an address in a single cycle (these operations are performed in a single cycle in the R2000 processor). The additional overhead may motivate another approach to address translation, such as accessing the data cache with virtual addresses rather than physical addresses. However, we will not examine such considerations here.

8.1 ORDERING OF LOADS AND STORES

It is very difficult to improve the performance of loads and stores by issuing them out of order, except that loads can be issued out of order with respect to stores if the out-of-order loads check for data dependencies on previous, pending stores. As we have already seen, stores can be performed only after all previous instructions have completed successfully. This preserves the in-order state in the data cache and external memory and permits recovery.

Checking for a memory dependency is more difficult than checking for a register dependency, because memory identifiers are larger than register identifiers. For example, the R2000 architecture has a 32-bit memory address and a 5-bit register number. In addition, we do not have much flexibility in checking memory dependencies, because dependencies can be detected only after address computation and translation have been performed, in contrast to register dependency checking which can occur during the decode stage.

In this section, we will consider various ways of relaxing ordering constraints on loads and stores while maintaining correct operation. We will use the following abstract code sequence to illustrate the operation of the various alternatives. In this sequence, the letters v, w, and x refer to program variables:

```
STORE   v           (1)
ADD                 (2)
LOAD    w           (3)
LOAD    x           (4)
LOAD    v           (5)
ADD                 (6)
STORE   w           (7)
```

In this instruction sequence, the *LOAD v* instruction (line 5) depends on the value stored by the *STORE v* instruction (line 1). Technically, the *STORE w* instruction (line 7) antidepends on the instruction *LOAD w* (line 3), but, because stores are held until all previous instructions are complete, antidependencies on memory locations are not of concern–nor are output dependencies, because stores are always performed in order.

8.1.1 Total Ordering of Loads and Stores

The simplest method of issuing loads and stores is to keep them totally in order with respect to each other–though not necessarily with respect to other instructions–as shown in Figure 8-1. In this case, there is no reason to have a separate reservation station for loads and stores, and these instructions can share a common reservation station. In Figure 8-1, the *STORE v* instruction is released in cycle 4 by the execution of all previous instructions. A store is released for issue when the corresponding entry of the reorder buffer reaches the writeback stage at the bottom of the reorder buffer. Because loads and stores are issued strictly in order to the same functional unit (the data cache), there is no need to detect dependencies between loads and stores. The disadvantage of this organization is that loads are held for stores, and stores in turn are held for all previous instructions. This causes memory accesses to have performance that is even worse than with in-order issue, because of the delay involved in releasing the store via the reorder buffer. This approach also causes the decoder to stall more often than with other alternatives described shortly, because the issue rate of loads and stores is constrained, causing the load/store reservation station to be full more often.

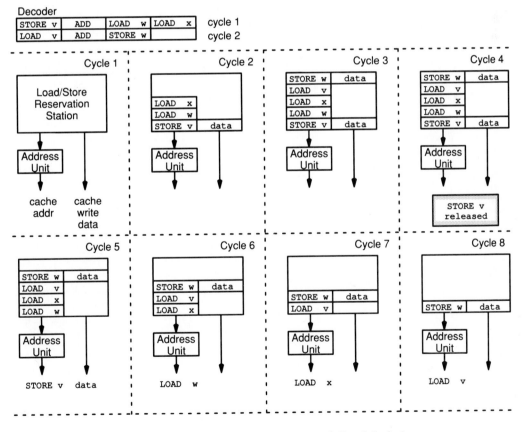

Figure 8-1. Issuing Loads and Stores with Total Ordering

8.1.2 Load Bypassing of Stores

To overcome the performance limitations of total ordering of loads and stores, loads can be allowed to bypass stores as illustrated in Figure 8-2. In this example, all loads but one are issued before the *STORE v* instruction is released for issue, even though

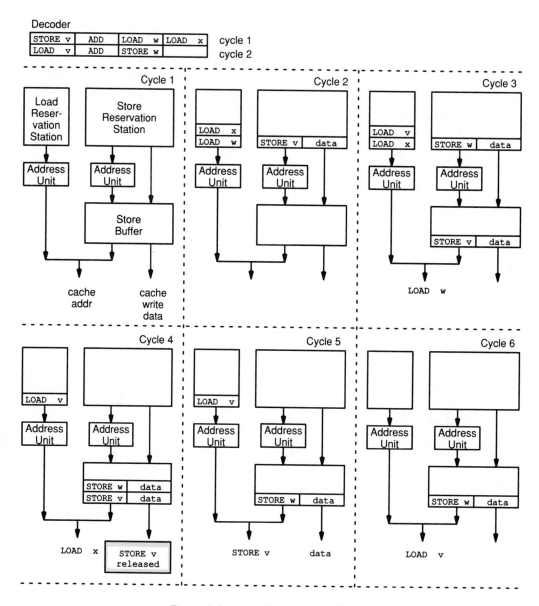

Figure 8-2. Load Bypassing of Stores

the loads follow the *STORE v* in the instruction sequence. Compared to Figure 8-1, load bypassing saves two cycles. To support load bypassing, the hardware organization shown in Figure 8-2 includes two, separate reservation stations and address units for loads and stores (Section 8.2 describes an organization that eliminates this duplication of hardware). The hardware also includes a store buffer.

The store buffer is required because a store instruction can be in one of two different stages of issue: before address computation and translation have been performed, and after address computation and translation have been performed but before the store is released and issued to the data cache. Store addresses must be computed before loads are issued so that load dependencies can be checked, but stores must then be held after address computation until all previous instructions complete. Figure 8-2 shows that the dependent *LOAD v* instruction is not issued until the *STORE v* instruction is issued, even though preceding loads are allowed to bypass this store. If a store address cannot be determined (for example, because its address-register value is not yet available), all subsequent loads are held until the address is valid. In this case, there might be a memory dependency, but the hardware cannot determine whether or not there is a dependency. The hardware must assume that the load is dependent until the hardware can check the dependency.

8.1.3 Load Bypassing with Forwarding

When load bypassing is implemented, it is unnecessary to hold a dependent load until the previous store is issued. To the contrary, as illustrated in Figure 8-3, the load can be satisfied directly from the store buffer if the address is valid and the data is available in the store buffer. This optimization, called *load forwarding,* avoids the additional latency caused by holding the load until the store is issued, as well as the latency caused by loading from the cache. With reference to Figure 8-2, load forwarding allows the *LOAD v* instruction to be satisfied three cycles earlier than without load forwarding (this instruction is issued one cycle earlier, and forwarding avoids the two-cycle latency of the data cache, making for a total benefit of three cycles).

Though it might seem strange for the processor to store a value that it will soon load back into a register, this behavior is sometimes necessary. For example, a compiler cannot always insure that an assignment to a variable (say, a parameter passed to a procedure) does not also modify another variable in the program (say, a global variable also accessed by the procedure). In this situation, the compiler cannot keep the potentially modified variable in a register while the assignment is performed. When variables might be *aliased,* so that two variables refer to an identical value, the compiler may generate frequent loads of values recently stored. Forwarding load data can help performance in this case.

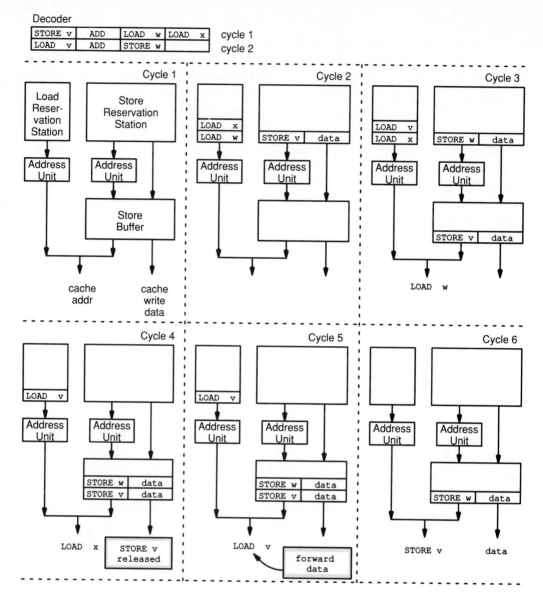

Figure 8-3. Load Bypassing of Stores with Forwarding

8.1.4 Performance of the Load/Store Policy

Figure 8-4 shows the performance of the various load/store mechanisms described in this section, for both a two-instruction and a four-instruction decoder. The interpretation of the chart labels is as follows:

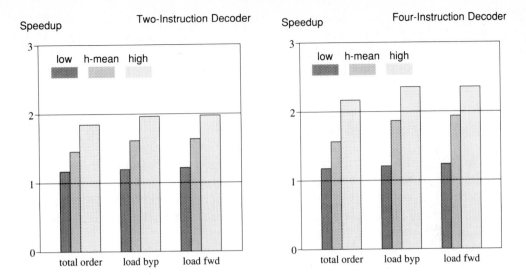

Figure 8-4. Performance of Various Load/Store Techniques

- total order – total ordering of loads and stores.
- load byp – load bypassing of stores.
- load fwd – load bypassing of stores with forwarding.

Load bypassing yields an appreciable performance improvement over total ordering: 11% with a two-instruction decoder and 19% with a four-instruction decoder. Load forwarding yields a much smaller improvement over load bypassing: 1% with a two-instruction decoder and 4% with a four-instruction decoder. The obvious conclusion is that the load/store mechanism should support load bypassing. Load forwarding is a less important optimization that may or may not be justified, depending on implementation costs.

8.1.5 Load Side Effects

In some systems, it is not possible to guarantee that loads are free of side effects. This is particularly true in systems that use memory-mapped input/output (I/O). Some I/O devices change state when accessed by a read; for example, some FIFO buffers sequence to the next data item and some device-status registers clear themselves. In this type of system, load bypassing is a dangerous operation. A bypassed load may be issued incorrectly, because of a mispredicted branch or exception. The bypassed load cannot be allowed to modify the system state incorrectly.

We can skirt this problem by noting that our processor includes a data cache and must already be aware that some types of data cannot be stored in the cache. External read accesses that change system state are a subset of *noncacheable* accesses. Rather than avoiding load bypassing altogether, we can have the processor prevent the bypassing only of noncacheable loads. This permits most load operations to take advantage of bypassing, without causing incorrect operation for the occasional noncacheable load.

8.2 ADDRESSING AND DEPENDENCIES

To implement load bypassing and load forwarding in the previous section, we used separate reservation stations and addressing units for loads and stores. This organization is flexible, conceptually straightforward, and minimizes the differences between the load and store functional units and other functional units. But it is an inconvenient implementation for two reasons.

First, this organization assumes two separate address units, each consisting of a 32-bit adder and (possibly) address-translation hardware. Having two address units seems unnecessary, because the issue rate of loads and stores is constrained by the data-cache interface to one access per cycle.

Second, this organization makes it difficult to determine the original order of loads and stores so that dependency checking can be performed. For example, when a load is issued, it may have to be checked against a store in the store reservation station, in the store address unit, or in the store buffer. Stores in the store reservation station have not had address computation and translation performed. If a load sequentially follows a store that is still in the store reservation station, the load must be held at least until the store is placed into the store buffer. However, there is no information in the reservation stations or store buffer, as defined, to allow the hardware to determine the set of stores which may cause dependencies for a given load. If the correct set of stores cannot be determined, the processor may unnecessarily hold loads because of dependencies, and load forwarding is not possible. Furthermore, deadlock is possible. A store may be holding because of a register dependency on a previous load, while the load is holding because it is considered (incorrectly) to have a memory dependency on the store. If this occurs, the processor cannot issue either instruction and will eventually halt.

In this section, we will examine an implementation of loads and stores that is more convenient than the one used in the previous section.

8.2.1 Limiting Address Logic with a Preaddress Buffer or Central Instruction Window

The hardware in Figure 8-2 can be reorganized as shown in Figure 8-5. This organization is based on a common window for loads and stores, similar to that shown in Figure 8-1. The window might be a combined reservation station for loads and stores, which we will call a *preaddress buffer*. The preaddress buffer is effectively a

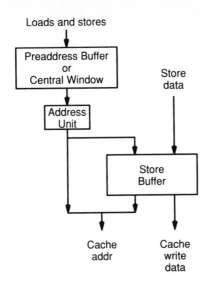

Figure 8-5. Reorganizing Load/Store Unit with a Preaddress Buffer or Central Window

reservation station for load and store addressing operations: the term "preaddress buffer" emphasizes that it is for both loads and stores and that it handles only the addressing operations for these instructions. Alternatively, the preaddress buffer may be replaced by a central window containing both loads and stores because it contains all instructions. In either case, loads and stores are issued to the address unit from the common window. A store buffer operates as before to hold stores until they are released and issued to the data cache.

When a load or store instruction is decoded, its address-register value (or a tag) and address offset are placed into the common window. The data for a store (or a tag) is placed directly into the store buffer. From the window, loads and stores are issued in order to the address unit (this can be accomplished with a central window using sequencing tags, as suggested in Section 7.2.3). A load or store cannot be issued from the window until its address register is valid, as usual. At the output of the address unit, a store address is placed into the store buffer, and a load address is sent directly to the data cache; thus loads bypass stores in the store buffer. Loads are checked for dependencies on stores in the store buffer. If a dependency exists, load data can be forwarded directly from the store buffer.

Figure 8-6 repeats the presentation of Figure 8-4 for the organization using a preaddress buffer or central window (there is no measurable performance difference between the preaddress buffer and central window). The results for totally ordered loads and stores are shown in Figure 8-4 for continuity, but having a common win-

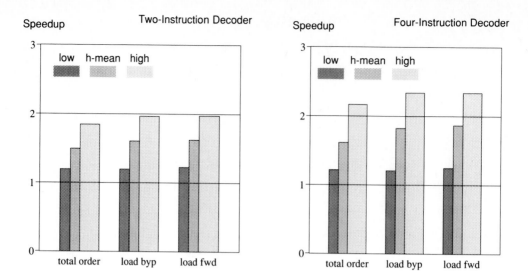

Figure 8-6. Performance of Various Load/Store Techniques Using a Common Window

dow and single address unit does not make a difference in this case. For these results, an eight-entry window and an eight-entry store buffer were used. In most cases, there is only a slight performance penalty caused by issuing loads to the address unit in order with stores: 1% with a two-instruction decoder and 3% with a four-instruction decoder. For the results in Figure 8-4, a load could be issued to the data cache while a store was simultaneously placed into the store buffer. Since this latter technique requires two address units, and since the common-window approach does not reduce performance very much, the implementation using the common window and a single address unit is a good cost/performance trade-off.

8.2.2 Effect of Store-Buffer Size

Figure 8-7 shows the effect of store-buffer size on performance, for a processor having an eight-entry preaddress buffer or central window and implementing load bypassing with forwarding. With a two-instruction decoder, both a four-entry and a two-entry store buffer have nearly identical performance to an eight-entry buffer. With a four-instruction decoder, a four-entry buffer causes a 1% performance loss over an eight-entry buffer, and a two-entry buffer causes a 4% loss. The four-entry buffer is a good choice for the four-instruction decoder, because it has good performance and the smaller number of entries facilitates dependency checking.

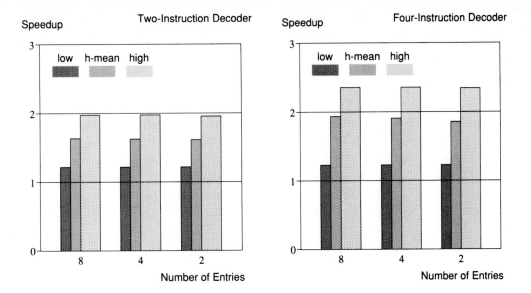

Figure 8-7. Performance Effect of Store-Buffer Size

8.2.3 Memory Dependency Checking

Using a common window and a single address unit greatly simplifies memory dependency checking, in comparison to the organization using separate load and store units in Figure 8-2, because loads and stores are issued to the address unit in order. When a load is issued to the data cache, its address is checked for dependencies against all active addresses in the store buffer. The store buffer contains only stores that sequentially preceded the load, and the store buffer is the only location of these stores until they are issued, after which dependency checking no longer matters.

If a load address matches an address in the store buffer and load forwarding is implemented, the load data is supplied, if available, directly from the store buffer. The dependency logic must identify the most recent matching store-buffer entry, because the address of more than one store-buffer entry may match the load address. The most recent entry contains the required data (or a tag for these data). If forwarding is not implemented, the load is held if its address matches the address of any store-buffer entry, without regard to the number or order of matching entries, until all matching store-buffer entries are issued to the data cache.

To reduce the effect of logic delays involved in dependency checking and forwarding, these operations can be performed as a load is issued to the data cache, with the load being canceled as required if a dependency is detected. Canceling may require support in the data cache and requires that the load cannot affect the state of the system before it can be canceled.

If load forwarding is not implemented, the comparators used to detect memory dependencies can be reduced by checking a subset of the full address. Figure 8-8 shows how performance is reduced as dependency checking is limited to eight, four, two, and one address bits. Although limiting address checking in this manner sometimes causes false dependencies, the effect on performance is very small until less than eight bits are used. For these measurements, low-order address bits were compared under the assumption that these bits have the most variation and are least likely to compare falsely. These results indicate that eight-bit comparators cause negligible reduction in performance–about .5% with a four-instruction decoder. However, some performance is lost because load forwarding is not possible with this technique, and this performance loss is not reflected in Figure 8-8.

If both load forwarding and address translation are implemented, the size of the dependency checking comparators can be reduced using information from the address-translation operation. If the address-translation system does not allow a program to map two different virtual addresses to the same physical location, the identifier of the translation lookaside buffer (TLB) entry used to translate an address provides a shortened form of the most significant bits of the address, and this identifier can replace these address bits in dependency checking. Under these conditions, comparisons involving the shortened addresses are exact. As an example of the savings, a 32-entry TLB and 4-Kbyte virtual-page size allow the dependency comparators to be reduced to 17 bits from 32 bits.

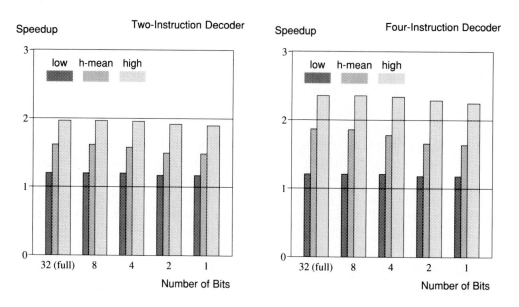

Figure 8-8. *Effect of Limiting Address Bits for Memory-Dependency Checking*

8.3 WHAT IS MORE LOAD/STORE PARALLELISM WORTH?

On the average, loads and stores are the second most frequent processor operations, after integer ALU operations. We concluded in Section 3.3.3, based on preliminary data, that having two data-cache interfaces was not worth the cost, even though an additional interface permits parallel operation of loads and stores, much as an additional ALU permits parallel operation of arithmetic and logical operations. It is illuminating to revisit that conclusion, now that we have refined the entire processor model.

We can obtain some parallelism between loads and stores without going so far as duplicating the data-cache interface. The data cache is the source of a very long-latency operation–processing a cache miss–and the cache can process a cache miss in parallel with other accesses without the expense of two cache ports. Though we have not discussed this detail, the standard processor model stalls the issuing of all loads and stores when a cache miss occurs, and resumes the issuing of loads and stores once the missing block has been reloaded into the cache. Since the processor allows out-of-order completion, it is relatively easy to remove this stall and permit the processor to issue loads and stores while the cache miss is being processed, as long as these other loads and stores do not also cause cache misses.

The processor is not concerned with the processing of a data-cache miss unless the miss occurs on a load, in which case the processor requires the data. The only additional cost to overlapping cache misses with other accesses is that the cache must be aware of result tags, in order to distinguish data from the overlapped reload from other data sent to the processor. Tags could be kept by the processor otherwise. When the cache has reloaded the missing cache block (this may not occur for a store miss), the cache simply returns the requested data to the processor along with its result tag. This technique can compensate somewhat for the reload penalty. However, even without this optimization, the superscalar processor still compensates for the reload penalty to some extent by executing instructions other than loads or stores during cache misses.

Beyond the preceding simple technique, we have one more option to increase load and store parallelism before duplicating the entire data-cache interface. This is an option that we have already seen: providing two reservation stations, one for loads and one for stores, with separate address units. This allows the processor to compute load addresses at the same time as store addresses. Of course, loads and stores still contend for the data-cache interface, and the processor can issue only one load or store per cycle. The processor cannot issue two accesses per cycle without an additional data-cache port.

Figure 8-9 shows the effect on performance of various techniques that increase the parallelism of loads and stores, with both a two-instruction and a four-instruction

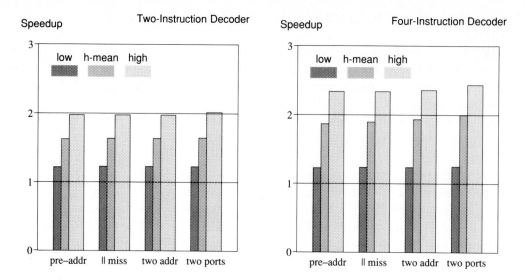

Figure 8-9. Performance Effects of Increasing Load and Store Parallelism by Various Techniques

decoder. In each case, load bypassing with forwarding is used. The interpretation of the chart labels is as follows:

- preaddr – a hardware organization using a preaddress buffer or central window and one address unit. This performance is equivalent to that of the "load fwd" case in Figure 8-6.

- ‖ miss – a hardware organization as in the "preaddr" case, except that the cache performs loads and stores in parallel with cache misses.

- two addr – a hardware organization using separate reservation stations and address units for loads and stores. This performance is equivalent to that of the "load fwd" case in Figure 8-4.

- two ports – a hardware organization using two data-cache ports and separate reservation stations and address units for loads and stores. Either data-cache port can be used for either a load or a store: a port is not dedicated to either loads or stores as the reservation stations and address units are.

The benefit of processing a cache miss in parallel with other operations is .6% with a two-instruction decoder and about 2% with a four-instruction decoder. Though this technique is simple, it does not have much benefit.

With a two-instruction decoder, there is a 1% benefit from duplicating the cache interface, but most of this is due to having two address units. In other words, the processor could realize most of the performance benefit without duplicating the data-cache interface. With a four-instruction decoder, there is a 7% performance benefit for duplicating the data-cache interface, but over half of this (4%) is due to having two address units. Two address units are expensive, both in terms of the complexity of memory-dependency checking and in terms of the hardware required—especially if the address units implement address translation. However, this expense may be justifiable, because it is localized on the processor chip and yields most of the benefit of load/store parallelism because it reduces the latency of loads.

On the other hand, the fact that most of the benefit of parallelism is due to two address units makes it impossible to justify duplicating the data-cache interface. Maintaining an issue rate of two load or store operations per cycle would require two sets of address-computation hardware (and, possibly, two sets of address-translation hardware), up to four data-cache buses (two for addresses and two for data), and large, duplicated arrays (or two-port arrays) to implement the data-cache tags and the data cache itself. Alternatively, this issue rate could be achieved by operating the data cache and data-cache interface at twice the processor's operating frequency, but we must reject this alternative because the data-cache interface is likely a severe performance impediment even when we try to operate it at the processor's operating frequency, much less twice this frequency.

8.4 ESOTERICA: MULTIPROCESSING CONSIDERATIONS

In a multiprocessing system, separate processors access shared data in order to cooperate on computing tasks. The requirement that processors communicate via shared data can constrain the issuing of loads and stores because of considerations beyond those we have already examined. This section describes how multiprocessing might affect the design of the load/store unit. Because our primary focus throughout this book is on the design of the processor rather than the system, this section may not interest all readers. However, we cannot ignore this topic completely, because the processor design has important implications on the design and capabilities of the multiprocessing system. It is important to weigh the performance advantage of a particular approach to loads and stores against the possible impact on the operation of the system.

The first consideration for a multiprocessing system deals with the ordering of loads and stores to shared data performed by more than one processor. For example, consider two processors that perform the following sequences of instructions:

Processor 1	Processor 2
STORE v	STORE w
LOAD w	LOAD v

In this sequence, each processor stores a value into a variable v or w, then loads the value stored by the other processor. Because the instructions are executed by different processors, there are several ways that the instructions of *processor 1* and *processor 2* might be positioned in time relative to each other. The instructions of both processors would execute in one of the following sequences:

Sequence A	Sequence B	Sequence C	Sequence D	Sequence E	Sequence F
1:STORE v	1:STORE v	1:STORE v	2:STORE w	2:STORE w	2:STORE w
1:LOAD w	2:STORE w	2:STORE w	1:STORE v	1:STORE v	2:LOAD v
2:STORE w	1:LOAD w	2:LOAD v	1:LOAD w	2:LOAD v	1:STORE v
2:LOAD v	2:LOAD v	1:LOAD w	2:LOAD v	1:LOAD w	1:LOAD w

These sequences all have something in common: at least one processor loads the value stored by the other processor. Or, at least, that is what we would naturally expect. However, we have examined some implementations of loads and stores in the superscalar processor that do not have this property. If the processors perform load bypassing, both processors might perform the loads before the stores, and neither processor would obtain the data written by the other. A multiprocessing system that behaves according to one of the foregoing sequences is said to provide *strong ordering* [Dubois et al. 1986]. Strong ordering simplifies the design of the system software, because it allows concurrent programs to behave in a way that we would naturally expect interleaved, sequential programs to behave. Since load bypassing issues loads out of order with respect to stores, strong ordering of memory accesses is difficult to achieve, and we might have to accept *weak ordering* of accesses, which imposes strong ordering only at the points of process interlocks. Strong ordering motivates a design that does not allow load bypassing of shared data.

The correct operation of a multiprocessing system also relies on *process interlocks*, such as *semaphores* [Dijkstra 1965], so that, for example, a program executing on one processor can exclude all other programs on all other processors from accessing shared data. Even if the system is designed for weak ordering of most accesses, strong ordering is needed for the correct operation of process interlocks. A process interlock can be set only before any access to the exclusive data–either a load or store access–and can be released only after all accesses to the shared data are complete. Otherwise, the process interlock cannot guarantee exclusive access to the shared data. Process-interlock operations are generally infrequent, though. Serialization of process-interlock operations (see Section 6.1.7) can insure that these operations are correctly performed, without reducing processor performance very much.

8.5 ACCESSING EXTERNAL DATA: OBSERVATIONS

Of all the alternatives examined in this chapter, an organization implementing load bypassing, with a common window for loads and stores and a store buffer, provides the best trade-off of hardware complexity and performance. This organization requires a single address unit that is used by loads and stores in their original program order. In the simplest organization, without load forwarding, memory dependency checking can be performed using the eight low-order bits of the memory addresses with no appreciable loss in performance. In this case, dependency checking is not concerned with the order of stores in the store buffer as it is when load forwarding is implemented. In any case, the store buffer requires only four entries. The performance with this organization is not the best that can be achieved, but is still good because it allows loads to be issued out of order with respect to stores.

We must conclude that the load/store interface of the superscalar processor is not really very difficult to implement, compared to other components of the processor (however, we have ignored CISC operations, such as register-to-memory operations, that would significantly complicate the operation of this interface). This conclusion possibly is a little surprising, given the nature of other parallel machines, such as array processors, which require very aggressive memory interfaces. However, we are trying to exploit instruction parallelism, rather than data parallelism. The benchmark programs we are using do not have a very high frequency of loads and stores compared to other instructions. Most computations involve data flowing to and from registers, rather than to and from memory, and thus performance is much more sensitive to register dataflow than to memory dataflow.

It is fortunate that the load/store unit is relatively simple, because it would be very difficult and expensive to achieve parallel operation of loads and stores. To emphasize a point made previously, superscalar processors have an advantage that they increase the utilization of expensive processor resources–the data cache and interface–without increasing the cost of these resources very much. Being used more frequently, the data cache and interface contribute more to performance, in relation to their cost, than in scalar processors.

Chapter 9

Complexity and Controversy

At this point, we have done a pretty thorough investigation of superscalar hardware, using an experimental approach to design that evaluates processor features in light of the performance they provide. Trace-driven simulation has proven a very useful tool in helping us avoid speculation about which processor features might be helpful and which might not. Simulation has permitted us to evaluate the performance of a large sample of general-purpose programs on a large number of superscalar design alternatives, without having to make estimates or idealizations about the behavior of either the programs or the processor components. We have identified major hardware features that are required for best performance and have uncovered techniques that simplify hardware with little reduction in performance. By defining both the potential and the limits of instruction parallelism, we have avoided either underdesigning or overdesigning the superscalar processor.

But some readers may be less than satisfied by our investigation–particularly those swayed, as I am, by the "RISC philosophy," which places great emphasis on simple hardware. We have seen that a general-purpose superscalar processor needs four major hardware features for best performance: out-of-order issue, register renaming, branch prediction, and a four-instruction decoder. However, I probably have not conveyed a good sense of the hardware complexity of these features in preceding chapters, discussing many hardware operations in rather abstract terms in order to aid understanding. This chapter delves a little more deeply into the implementation to provide a better understanding of the complexity. These features verge on being unacceptably complex, and their complexity may argue against the design goal of attempting to achieve the best possible performance.

The complexity of superscalar hardware motivates us to look at simplifications. We will see that the four major hardware features just listed–out-of-order issue, register renaming, branch prediction, and a four-instruction decoder–are fundamental hardware techniques for performance. We cannot eliminate these without a big detriment to performance, and there are no simple hardware alternatives that provide nearly the same performance. However, we have, throughout the course of our investigation, identified several incremental simplifications that do not reduce performance very much. In this chapter, we will see what happens when we implement a number of these simplifications at the same time, something we have not considered before. The cumulative effect of these simplifications might reduce performance by an unacceptable amount, but we will see that this is not much of a problem.

The hardware complexity also motivates us to look for ways in which software can aid the hardware design. This chapter sets the stage for an examination of software superscalar techniques in the following two chapters.

9.1 A BRIEF GLIMPSE AT DESIGN COMPLEXITY

Previous chapters have argued that a reorder buffer should be used to implement recovery and renaming and have argued that forwarding is required to sustain performance. This section considers, in detail, a straightforward implementation of the instruction decoder, reorder buffer, reservation stations, and forwarding logic in light of the requirements of register dataflow and recovery. The implementation presented here provides a good illustration of the hardware complexity, showing that the best hardware alternative is quite complex.

9.1.1 Allocating Processor Resources

During instruction decode, the processor allocates result tags, reorder-buffer entries, and reservation-station entries to the instructions being decoded. The processor allocates a resource to an instruction by providing an identifier for the resource, using a hardware structure as shown in Figure 9-1. Hardware similar to that shown in Figure 9-1 is used for each resource to be allocated; for example, a version of this hardware is required for each reservation station, to allocate entries to instructions.

The allocation hardware has multiple stages, with one stage per instruction (these stages are not pipelined–they all operate in a single cycle). This organization takes advantage of the fact that resource identifiers are small (for example, a 16-entry reorder buffer requires a 4-bit entry identifier) and also takes advantage of the fact that allocation has nearly a full processor cycle to complete. The input to the first stage is an identifier for the first available resource. If the first instruction requires the resource, the first allocation stage uses this identifier and forms an identifier for the next available resource that is passed on to the second stage. If the first instruction does not require the resource, the identifier is passed on to the second stage unmodified. For example, in the case of reorder-buffer allocation, the first instruction would receive the identifier for the first available reorder-buffer entry, the second

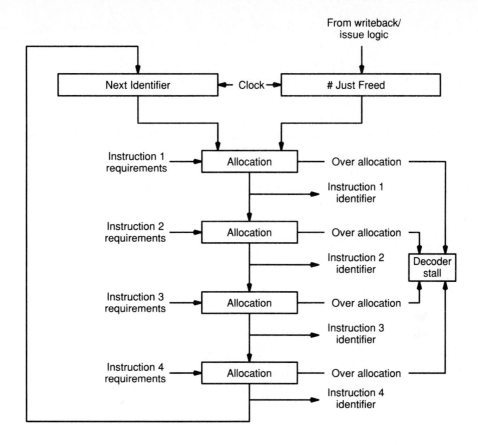

*Figure 9-1. Allocation of Result Tags, Reorder-Buffer Entries,
and Reservation-Station Entries*

instruction would receive the identifier for the second available reorder-buffer entry, and so on.

The identifier for the first available resource is based upon the resources allocated and freed in the previous cycle. For example, again in the case of reorder-buffer allocation, assume that the fifth reorder-buffer entry was allocated to the final instruction in the previous decode cycle and that two reorder-buffer entries were written into the register file, thus freed. In the current cycle, the last instruction of the previous decode cycle would be in the third reorder-buffer entry, because the reorder buffer has been advanced by two entries. The first valid instruction in the current decode cycle would be allocated the fourth entry, and other instructions would be allocated the fifth and subsequent entries.

Each allocation stage–other than the first one–is essentially an incrementer for resource identifiers. The stage either increments the identifier or passes it unmodified, depending on the needs of the corresponding instruction. If any incrementer attempts to allocate beyond the last available resource, the decoder is stalled. The stalled instruction, and all subsequent instructions, repeat the decode process on the next cycle. The incrementers need not generate identifiers in numerical sequence, because the identifiers are used by hardware alone. The identifiers can be defined so that the incrementers take minimal logic.

In this implementation, we assume that there is separate hardware to allocate result tags for instructions. Reorder-buffer identifiers are inappropriate to use as result tags, because these identifiers are not unique. Reorder-buffer entries are allocated at the top of a FIFO, and it is likely that the same entries will be allocated over and over again as entries are removed from the FIFO. However, the result tags do have the same number of bits as reorder-buffer identifiers, because both keep track of pending updates to registers.

9.1.2 Instruction Decode

During instruction decode, the dependency logic must insure that all instructions receive operands or tags. In addition, the state of the dependency hardware must change to reflect the new instructions, so that subsequent instructions correctly obtain operands. Figure 9-2 illustrates some of the dependency operations performed during instruction decode, focusing on the operations performed in the reorder buffer.

To obtain operands, the reorder buffer is associatively searched using the source-register identifiers of the decoded instructions. Since Figure 9-2 shows four reorder-buffer ports for a four-instruction decoder, the source-register identifiers are selected by a register-port arbiter. These source-register identifiers are compared to result-register identifiers of previous instructions stored in the reorder buffer, at each of four read ports. If the result is not available, a result tag is obtained. As Figure 9-2 shows, result tags use the same storage as result values, so that operand values and tags are treated uniformly and use the same buses for distribution. This minor optimization does not handle all requirements for tag distribution, as discussed shortly. If instructions arbitrate for register ports and thereby operand buses, as assumed here, the reservation stations must match instructions to operands or tags, so that these can be stored properly.

Supplying operands to reservation stations requires two paths that are omitted from Figure 9-2 for clarity. The first of these paths distributes instruction constants, such as offsets for memory addresses, from the decoder to the reservation stations. The second path distributes tags for instructions that have dependencies on instructions decoded in the same cycle. These tags are supplied by the tag-allocation hardware, rather than the reorder buffer. Detecting dependencies between instructions

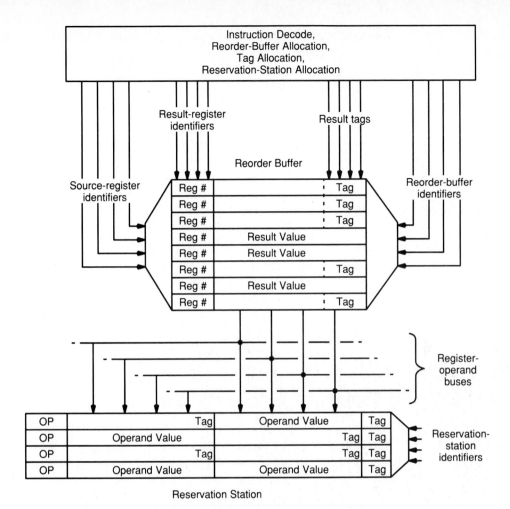

Figure 9-2. Instruction Decode Stage

decoded at the same time must be based on comparisons between source-register and destination-register identifiers of decoded instructions, taking 12 comparators. In either case, the dependent operands are not allocated operand buses, and thus require other buses to distribute tags to the reservation stations.

There is a potential pipeline hazard in the forwarding logic that the design must avoid [Weiss and Smith 1984]. This hazard arises when a decoded instruction obtains an operand tag for a value that will appear on a result bus in the next cycle. In this case, the operand tag obtained by the instruction is placed into the reservation station too late to be compared with the result tag that appears on a result bus during the decode cycle (recall that a result tag appears a cycle before the actual result). The instruction misses the forwarded result, causing incorrect operation. Avoiding this hazard takes eight additional comparators to detect, during decode and independent of the reservation stations, that forwarding is required in the next cycle.

At the end of the decode cycle, the result-register identifiers and result tags of decoded instructions are written into the reorder buffer. These should be written into the empty locations nearest the bottom of the buffer, as identified by the allocation hardware. For a four-instruction decoder, this requires four write ports, at each location of the reorder buffer, on the storage for register identifiers and tags. The four write ports could be avoided by allocating entries only at the top of the reorder buffer, in which case only the entries at the top would require four write ports. However, allocating entries at the top of the reorder buffer complicates the associative lookup, because it makes it difficult to determine the location of valid entries; there can be invalid entries among valid ones if only fixed locations are allocated. This also increases the likelihood that the reorder buffer will not have a sufficient number of entries, because the entries at the top of the reorder buffer must be cleared before the reorder buffer can accept new instructions. Finally, this approach increases the amount of time taken for a result to reach the bottom of the reorder buffer, increasing the branch delay if the entire reorder buffer is flushed on a mispredicted branch as discussed in Section 5.2.1.

9.1.3 Instruction Completion

When a functional unit produces an instruction result, the result is written into the reorder buffer after the functional unit successfully arbitrates for a result bus. The result tag identifies the reorder-buffer entry to be written. The reorder-buffer identifier that was allocated during decode was used only to write the result tag and the result-register identifier into the reorder buffer, and is meaningless in subsequent cycles because entries advance through the reorder buffer as results are written to the register file. Thus, the result write must be associative: the written entry is the one with a result tag that matches the result tag of the completed instruction. This is true of any reorder buffer operated as a FIFO, with or without a future file.

9.1.4 The Painful Truth

For brevity, we have examined only a small portion of the superscalar hardware: the portion of the decoder and issue logic that deals with renaming and forwarding. This is a very small part of the overall operation of the processor. We have not begun to consider the algorithms for maintaining branch-prediction information in the instruction cache and its effect on cache reload, the mechanisms for executing branches and checking predictions, the precise details of recovery and restart, and so on. But there is no point in belaboring the implementation details. We have seen quite enough evidence that the implementation is far from simple.

Just the hardware presented in this section requires 64 5-bit comparators in the reorder buffer for renaming operands: 32 4-bit comparators in the reorder buffer for associatively writing results: 60 4-bit comparators in the reservation stations for forwarding: logic for allocating result tags and reorder-buffer entries: and a reorder buffer with 4 read ports and 2 write ports on each entry, with 4 additional write ports on portions of each entry for writing register identifiers, results tags, and instruction status. The complexity of this hardware is set by the number of uncompleted instructions permitted, the width of the decoder, the requirement to restart after a mispredicted branch, and the requirement to forward results to waiting instructions.

If the trend in microprocessors is toward simplicity, we are certainly bucking that trend.

9.2 MAJOR HARDWARE FEATURES

The hardware complexity facing us is mostly caused by four major hardware features: out-of-order issue, register renaming, branch prediction, and a four-instruction decoder. To decide whether this complexity is justified, the first thing we should consider is the value of these features. Table 9-1 summarizes the value of each feature in terms of its performance advantage. Each entry in Table 9-1 is the relative increase in performance due to adding the given feature, in a processor that already has all other listed features. This simple summary indicates that each of these features is important in its own right. However, these features are interdependent in ways not illustrated by Table 9-1.

Branch prediction and a four-instruction decoder overcome the two major impediments to supplying an adequate instruction bandwidth for out-of-order issue: decoder stalls during the time that the processor determines the outcome of a branch, and decoder inefficiency caused by instruction misalignment. The techniques used to overcome the branch delay in a scalar processor, such as compiler scheduling of branch delays or predicting the outcome of a single branch, are not effective in a superscalar processor for general-purpose applications because of the frequency of branches and the magnitude of the branch-delay penalty. In a processor achieving a speedup of two, the branch delay represents about six to eight instructions. A four-instruction decoder increases the value of branch prediction more than does a two-

Table 9-1. Performance Advantage of Major Processor Features

Out-of-order issue	Register renaming	Branch prediction	Four-instruction decoder
52%	36%	30%	18%

---

instruction decoder, helping supply adequate instruction bandwidth even in the face of misaligned instruction runs.

Out-of-order issue and register renaming not only provide performance in the expected ways, by improving machine and instruction parallelism, but also provide performance by supporting branch prediction. The instruction window for out-of-order issue provides a buffer into which the instruction fetcher can store instructions following a predicted branch, and register renaming provides a mechanism for recovering from mispredicted branches.

Also, Table 9-1 implies another way in which out-of-order issue, register renaming, branch prediction, and a four-instruction decoder depend on each other. These features together provide more performance than is provided by each feature taken separately, because each relative improvement in Table 9-1 assumes that the other features are already implemented. Thus, it is difficult to justify implementing anything but the complete set of features, except that this observation presupposes that cycle time is not affected and that the performance goal justifies the design complexity.

The complexity of the superscalar processor is its most troubling aspect. None of the major features by itself is particularly difficult to implement, but interdependencies between these features create complex hardware and long logic delays. Complexity is significantly increased by the goal of decoding, issuing, and executing more than one instruction per cycle. Often, this complexity manifests itself in the arbitration logic that allocates multiple, shared processor resources among multiple contenders. The superscalar processor has a number of different arbiters, each with its own complexity and timing constraints. The design of these arbiters has been widely ignored in the published literature and in this book because arbiter design is very dependent on the hardware technology and implementation.

9.3 HARDWARE SIMPLIFICATIONS

Though a superscalar processor is complex, we have seen that many hardware simplifications are possible. We have examined the incremental effects of each simplification on the performance of a standard evaluation model. But this gives rise to a concern: it is possible that, when a superscalar processor is simplified in a number of ways, the cumulative effect of these simplifications reduces performance more than removing one of the major processor features listed in Section 9.2. If this were the case, we would be forced to conclude that even minor simplifications were not possible. Either this, or that the case for superscalar processors were even less hopeful. This section summarizes the cumulative effects of some of the simplifications we have examined, showing that this concern is not valid, at least to the extent that the cumulative effect is not as severe as removing out-of-order issue, register renaming, branch prediction, or a four-instruction decoder.

Table 9-2 lists several progressively simplified processor designs and the cumulative performance degradation. The cumulative degradation of each design is the percentage reduction in performance (based on the harmonic mean) relative to our standard processor configuration. Figure 9-3 and Figure 9-4 present the speedups of the configurations in Table 9-2. Although one conclusion presented in Section 9.2 was that a four-instruction decoder is desired over a two-instruction decoder, data for the two-instruction decoder is presented here for two reasons. First, this shows the effects of hardware simplifications in a processor that is more severely instruction-fetch limited than a processor with a four-instruction decoder. Second, this illustrates that a four-instruction decoder still achieves better performance than a two-instruction decoder even with simplified hardware.

Many of the hardware simplifications do not reduce performance very much because the instruction window decouples instruction-fetch and instruction-execution limitations. The average rates of fetching and execution are more important to overall performance than short-term limitations. Other simplifications are the direct result of the major hardware features described in Section 9.2. For example, the instruction-ordering information retained by the reorder buffer allows this information to be irrelevant in the instruction window, avoiding the need to compress the window

Table 9-2. Cumulative Effects of Hardware Simplifications

	Hardware Simplification (cumulative)	Degradation (cumulative)	
		2-inst	4-inst
a	execution model described in Section 3.4, with branch prediction in instruction cache	n/a[1]	n/a[2]
b	single-port array in instruction cache for branch-prediction information (Section 4.3.4)	−3%	−4%
c	limiting decoder to four register-file read ports (Section 4.4) and decoding a single branch instruction per cycle (Section 4.5.3)	−3%	−6%
d	waiting until a mispredicted branch reaches the bottom of the reorder buffer before restarting (Section 5.2.1)	−7%	−11%
e	central instruction window (not compressed—Section 7.2.3), limited operand buses (Section 7.2.4), single address unit (Section 8.2.1)	−8%	−15%
f	no load forwarding from store buffer (Section 8.1.2), 8-bit dependency checking (Section 8.2.3)	−9%	−17%
g	no load bypassing of stores (Section 8.1.1)	−16%	−28%

[1]Speedup = 1.64.   [2]Speedup = 1.94.

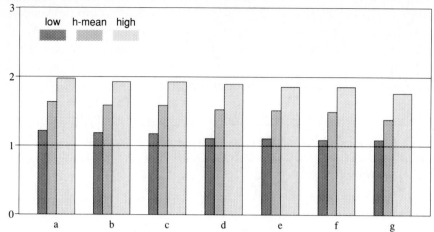

Figure 9-3. Cumulative Simplifications with Two-Instruction Decoder

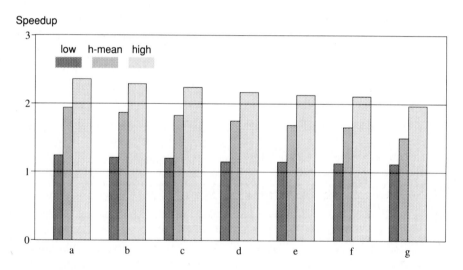

Figure 9-4. Cumulative Simplifications with Four-Instruction Decoder

and simplifying its implementation. As a further example, the ability of the instruction window to adapt dynamically to constraints allows good performance with a limited number of operand buses between the central window and the functional units.

Eliminating load bypassing is the only simplification listed in Table 9-2 that causes lower performance than the removal of out-of-order issue, register renaming, branch prediction, or a four-instruction decoder. Apparently, the conclusion of Section 8.1.4–that load bypassing is the most important feature of the load/store mechanism–is still valid in a simplified processor.

9.4 IS THE COMPLEXITY WORTH IT?

Though we might conclude from the foregoing sections that the hardware complexity of a superscalar processor is not completely unmanageable, the hardware still is complex. We also have not considered the effect of this complexity on the processor cycle time and design time. Though out-of-order issue, register renaming, and branch prediction have been implemented on other computers as early as the 1960s [Tomasulo 1967], we must be careful about inferring that these techniques are therefore easily adapted to microprocessors. Design errors in microprocessors which we might introduce with superscalar techniques would generally be a lot more difficult to find and correct than similar errors in board-level computers. And we have not considered the other part of the "RISC philosophy": that software techniques permit high performance with simple hardware–*especially* with simple hardware.

Our investigation already has suggested several areas where software support can improve performance or simplify the hardware. In Chapter 4, we saw that software can improve the efficiency of the instruction fetcher by aligning instruction runs and moving instructions after a branch to pad the decoder. Simulation of hardware techniques for aligning and merging instruction runs indicated that these can be important optimizations, but it is easier for software to provide these optimizations than it is for hardware to provide them. The suggested software optimizations are only two of many such optimizations that can either improve performance, simplify hardware, or both.

Our simulations have been based on code generated by a compiler that has no notion of optimal code for a superscalar processor. Software might be able to help us eliminate out-of-order issue by reordering instructions in the object code. If we can eliminate out-of-order issue, we can reduce the number of instructions that the processor has to consider at once, and we can eliminate the instruction window. Then, because there are not very many instructions to consider at once, the dependency logic is much simpler, even if we decide that software cannot completely eliminate the need for this dependency logic. Without out-of-order issue, renaming is not very important.

We have not considered in any depth what aid we might expect from software techniques. It is irresponsible to assume that a superscalar design is worthwhile on the evidence we have seen so far. So we will devote the rest of this book to an examination of software techniques and to the controversy between hardware-intensive and software-intensive approaches to superscalar design.

Chapter 10

Basic Software Scheduling

Software scheduling is the process of arranging the order of instructions in object code so that they are executed by hardware in an optimum (or near optimum) order. Software scheduling increases the likelihood that the processor fetches independent instructions and helps insure that the processor executes time-critical operations efficiently. Both of these increase resource utilization and instruction throughput.

This chapter shows how a software scheduler can arrange sequences of instructions for improved performance. It considers the problem of scheduling code sequences having the property that all instructions within the sequence execute together. Such sequences have the advantage that the software scheduler knows the dependencies between instructions in the sequence–because there are no branches or branch targets in the middle of the sequence. Even though all real programs contain branches, the scheduling of these sequences is an important software-scheduling task that forms the basis of more general scheduling. Chapter 11 explores techniques for scheduling code sequences that contain branches or branch targets within the sequence.

The ultimate purpose of this chapter is to examine an alternate view of superscalar design. This view is that the superscalar hardware should be simple, with software optimizations providing performance in lieu of the complex hardware we examined in preceding chapters. For now, however, we will not consider simple hardware. Rather, we will assume hardware that has all of the features we concluded were necessary–at least necessary without software support–in the previous chapter. This helps make the important point that software scheduling is useful even with complex hardware. After exploring the fundamental principles of software scheduling, we will return to the question of whether software optimizations can allow simple hardware.

10.1 THE BENEFIT OF SCHEDULING

To understand the benefit of software scheduling, consider Figure 10-1. A typical RISC compiler translates the C source-language statement into the assembly-language sequence shown (the Am29000™ instruction set [AMD 1989] is used for the purpose of illustration, with register designators replaced by symbolic names for clarity). Figure 10-1 also shows how this sequence of instructions is executed by a superscalar processor, assuming a processor configuration similar to the standard configuration presented in Chapter 3, and assuming that the code sequence is aligned in memory so that the first instruction in the sequence is in the first position of the decoder. Figure 10-1 shows the decoding of instructions in this sequence, the execution in the functional units, and the cycles in which instruction results are available

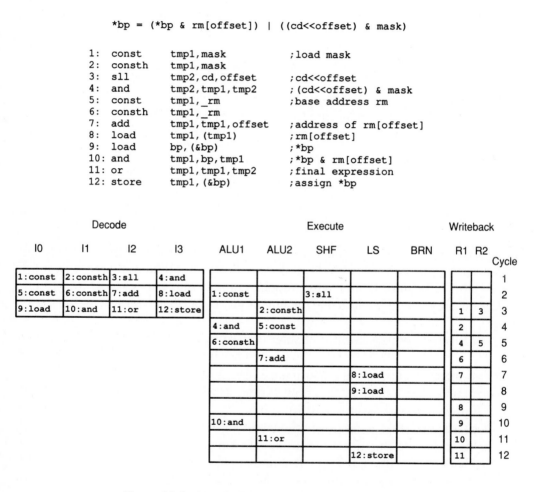

```
*bp = (*bp & rm[offset]) | ((cd<<offset) & mask)

 1: const    tmp1,mask         ;load mask
 2: consth   tmp1,mask
 3: sll      tmp2,cd,offset    ;cd<<offset
 4: and      tmp2,tmp1,tmp2    ;(cd<<offset) & mask
 5: const    tmp1,_rm          ;base address rm
 6: consth   tmp1,_rm
 7: add      tmp1,tmp1,offset  ;address of rm[offset]
 8: load     tmp1,(tmp1)       ;rm[offset]
 9: load     bp,(&bp)          ;*bp
10: and      tmp1,bp,tmp1      ;*bp & rm[offset]
11: or       tmp1,tmp1,tmp2    ;final expression
12: store    tmp1,(&bp)        ;assign *bp
```

| Decode | | | | Execute | | | | | Writeback | | |
I0	I1	I2	I3	ALU1	ALU2	SHF	LS	BRN	R1	R2	Cycle
1:const	2:consth	3:sll	4:and								1
5:const	6:consth	7:add	8:load	1:const		3:sll					2
9:load	10:and	11:or	12:store		2:consth				1	3	3
				4:and	5:const				2		4
				6:consth					4	5	5
					7:add				6		6
							8:load		7		7
							9:load				8
									8		9
				10:and					9		10
					11:or				10		11
							12:store		11		12

Figure 10-1. Unscheduled Code and Execution Timing

on the result buses. As shown, the sequence can be fetched and decoded in three cycles, but a total of 11 cycles are required for execution. (To simplify Figure 10-1, loads and stores are totally ordered. The *store* instruction at the end of the sequence is shown being issued at the load/store unit when all of its dependencies are removed. It should be understood that the *store* instruction might instead be issued to a store buffer as soon as its address could be computed.)

10.1.1   Impediments to Efficient Execution

The instruction sequence in Figure 10-1 has been translated by a right-to-left scan of the C source statement, with no attention to the execution of these instructions by the target hardware. This approach encounters two sources of inefficiency. The most severe problem is with the sequence of instructions 5 through 12, which take the longest to execute. Instruction 5 is not decoded until the second cycle–even though it is independent of preceding instructions–and does not begin execution until the fourth cycle. The reason for the decode-to-execution delay is that instruction 5 conflicts with instruction 2 for the functional unit *ALU2*. Furthermore, the independent *load* instruction 9 waits behind the dependent *load* instruction 8, delaying the time that instruction 10 and subsequent instructions can begin. Both these conflicts are unfortunate, because there are plenty of cycles in which the requisite functional units are available. The second problem is that potential concurrency in the code is not exposed, because the independent instruction sequences 1 through 4 and 5 through 8 are fetched in different cycles rather than being interleaved with each other. In the first two cycles, the processor fetches more dependent instructions than independent ones.

This example illustrates that instruction scheduling has several goals. The first goal is to start as early as possible those sequences of instructions that take the longest to execute. The second goal is to help avoid resource conflicts that unnecessarily increase execution time. The third goal is to help the hardware discover instructions that can be executed at the same time.

10.1.2   How Scheduling Can Help

By comparison to Figure 10-1, Figure 10-2 shows a functionally identical sequence of instructions that requires only 8 cycles to execute (the cycles labeled "–" are those in which the corresponding functional unit is stalled because of a result-bus conflict). The code sequence in Figure 10-2 differs from the sequence in Figure 10-1 in that the instructions have been scheduled. The advantage to scheduling is that the instruction sequence 5 through 12 is begun in the first execution cycle and does not encounter conflicts for functional units that increase execution time. Also, more independent operations are fetched in the first 2 cycles than in the unscheduled sequence. In this example, scheduling has saved 3 cycles out of 11 in the execution time of the code sequence.

```
*bp = (*bp & rm[offset]) | ((cd<<offset) & mask)
```

```
5:  const    tmp1,_rm              ;base address rm
6:  consth   tmp1,_rm
9:  load     bp,(&bp)              ;*bp
7:  add      tmp1,tmp1,offset      ;address of rm[offset]
1:  const    tmp2,mask             ;load mask
8:  load     tmp1,(tmp1)           ;rm[offset]
3:  sll      tmp3,cd,offset        ;cd<<offset
2:  consth   tmp2,mask
4:  and      tmp2,tmp2,tmp3        ;(cd<<offset) & mask
10: and      tmp1,bp,tmp1          ;*bp & rm[offset]
11: or       tmp1,tmp1,tmp2        ;final expression
12: store    tmp1,(&bp)            ;assign *bp
```

Decode				Execute					Writeback		
I0	I1	I2	I3	ALU1	ALU2	SHF	LS	BRN	R1	R2	Cycle
5:const	6:consth	9:load	7:add								1
1:const	8:load	3:sll	2:consth	5:const			9:load				2
4:and	10:and	11:or	12:store		6:consth	3:sll			5		3
				7:add	1:const		—		6	3	4
				2:consth	4:and		8:load		7	1	5
					—				9	2	6
				10:and					4	8	7
					11:or				10		8
							12:store		11		9

Figure 10-2. Scheduled Code and Execution Timing

Compilers for scalar RISC processors perform scheduling to avoid pipeline conflicts and dependencies [Auslander and Hopkins 1982, Hennessy and Gross 1983]. However, the goal of this pipeline scheduling is more modest than the goal of the superscalar scheduler. For example, a scalar pipeline scheduler would note that the result of the load instruction 9 in Figure 10-1 is used immediately by the next instruction, causing a one-cycle pipeline stall (or *no-op*, if there are no hardware interlocks). To avoid this additional cycle, the pipeline scheduler could move another, independent operation after the load to be performed concurrently with part of the load. This is the only possible benefit of pipeline scheduling in this sequence, because all other results are produced in a single cycle and are not needed by another instruction until the end of the cycle. In contrast, the superscalar scheduler is concerned with a much larger number of instructions that might be decoded before their results are available and a much larger number of potential resource conflicts, both of which complicate the task of the scheduler. At the same time, the potential benefit

Superscalar Microprocessor Design

of the superscalar scheduler is larger, because the overall instruction-execution rate is higher. For example, the scalar pipeline scheduler can save 1 cycle in 13–an 8% benefit–whereas the superscalar scheduler can save 3 cycles in 11–a 27% benefit.

10.1.3 Is the Benefit Significant?

The conclusions in Section 10.1.2 are on a shaky foundation, because they favor software scheduling on the basis of a single example. Moreover, this example is a large instruction sequence containing two independent expressions. It is reasonable to ask whether this example overstates the benefit of scheduling.

Chapter 4 showed that general-purpose applications have a large number of short instruction runs, limiting instruction throughput. There is another side to this observation. Long, unscheduled instruction runs also limit instruction throughput, but because they are execution limited rather than fetch limited. For example, the unscheduled sequence in Figure 10-1 has a throughput of 1.1 instructions per cycle (ignoring subsequent instructions that might be executed concurrently), whereas scheduling increases the instruction throughput to 1.5 instructions per cycle. Long instruction sequences are potentially a fruitful source of performance, because they are not instruction-fetch limited, and because they are more likely to contain independent operations than short sequences and thus more likely to benefit from scheduling.

However, the benefit of scheduling long instruction runs is proportional to the execution time of these long runs in unscheduled code. For example, if the execution time of long runs takes 75% of the overall execution time, and scheduling reduces the execution time of these runs by 20%, scheduling has a 15% benefit to the overall execution time. In a scalar processor, long instruction runs certainly tend to dominate overall execution time, simply because they are long. It is not very clear that this is also true in a superscalar processor. Compared to the execution times in a scalar processor, a superscalar processor probably reduces the execution time of long instruction runs more than it reduces the execution time of short instruction runs. Unfortunately, there is no clear method of quantifying the execution time of long instruction runs in a processor with out-of-order issue, because long instruction runs are overlapped with other runs. It is difficult to say how the execution time is really allocated among the runs, and thus it is difficult to say what effect scheduling has on the overall execution time without actually writing a software scheduler and measuring the benefit.

The benefits and trade-offs of scheduling discussed in this chapter and Chapter 11 are based on scheduling as performed for scalar RISC processors to avoid pipeline interlocks and for vector and VLIW processors to increase the number of concurrent operations. Though these scheduling problems are very similar to each other and to the scheduling problems we examine here [Hennessy and Gross 1983], the actual benefit of scheduling in a superscalar processor, and the interaction with hardware features such as renaming and out-of-order issue, have not been extensively studied [Lam 1990]. It is worthwhile to consider the issues surrounding software scheduling and how this scheduling can affect the processor architecture and its performance, especially since we want to determine whether or not software scheduling can allow the hardware to be simplified. But we should be careful about making general conclusions about the performance benefit, because we cannot measure the benefit.

10.2   PROGRAM INFORMATION NEEDED FOR SCHEDULING

Software scheduling depends on information that the scheduler extracts from the program. This section and Section 10.3 provide an introduction to the information gathered prior to and used during scheduling.

10.2.1   Dividing Code into Basic Blocks

We have been examining software scheduling applied to code sequences that do not contain a branch or a branch target within the sequence. The formal term for such a sequence is *basic block*. A basic block has the property that if any instruction in the basic block is executed, all other instructions are also executed. Since all instructions in the basic block execute together, the dependencies between them can be known precisely (except for dependencies on memory locations whose addresses are unknown to the scheduler). This property is what imposes the requirement that no branch or branch target can appear within the block.

Figure 10-3 shows an example program fragment partitioned into basic blocks. Figure 10-3 shows both the basic blocks and the *control flow graph* of this code. The control flow graph shows the possible paths of control flow through the basic blocks; the actual path taken through the graph during execution depends on the outcomes of the various branches at the end of some of the basic blocks. The algorithm for partitioning a program into basic blocks is straightforward [Aho et al. 1986]:

1.   Determine the set of all *leaders* in the program. A leader is the first instruction in the program, the target of any branch, or any instruction following a branching instruction.

2.   Partition the program into basic blocks, where each block contains a single leader and all instructions up to but not including the next leader.

An exception to these rules is possible. A subroutine call can appear within a basic block, even though it is effectively a branch operation. If a subroutine is required to return to the calling program–which is usually the case in procedural languages such as C–the call does not violate the restriction that all instructions in the basic block execute together. Including subroutine calls within basic blocks rather

```
A;
B;
do {
    D;
    E;
    if ( F ) {
        G;
        H;
    }
    else {
        I;
    }
    J;
    K;
    L;
} while ( C )
```

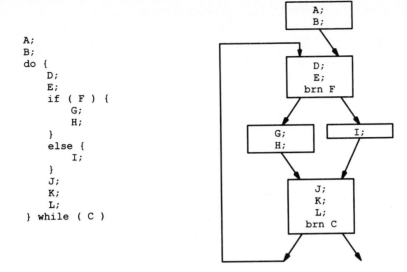

Figure 10-3. Partitioning Code into Basic Blocks

than having them end basic blocks has the advantage that basic blocks are generally longer, and the scheduler has a better chance of finding independent operations. However, since the subroutine can use and/or modify program data, the data dependencies arising from the call must be accounted for.

After the code is partitioned into basic blocks, the control flow graph is constructed by determining the target basic blocks of all branches. The *immediate successors* of a basic block are those blocks that can execute just after the block has executed (the *successors* of a basic block are any blocks that can execute after the block). Basic blocks that do not contain branches–or that contain unconditional branches–have one immediate successor. For example, in Figure 10-3, basic block *G-H* has one immediate successor. Basic blocks that contain a conditional branch normally have two immediate successors, because conventional branches have two possible outcomes: there is one immediate successor for the taken branch and another for the nontaken branch. For example, in Figure 10-3, basic block *D-E-brn F* has two immediate successors.

In the control flow graph, the *immediate predecessors* of a basic block are those blocks that can execute just before the block (the *predecessors* of a basic block are any blocks that can execute before the block). In Figure 10-3, basic blocks *G-H* and *I* are immediate predecessors of block *J-K-L-brn C*. A block can have many immediate predecessors. Also, within a loop, a block can be its own successor and predecessor: this is true for every basic block within the loop in Figure 10-3.

10.2.2 The Dataflow Graph of a Basic Block

Once a program has been partitioned into basic blocks, a dataflow graph can be constructed for each basic block. The dataflow graph represents the production and use of data within the basic block and thus specifies true dependencies that must be satisfied by the instruction scheduler. Information derived from the dataflow graph aids the scheduler in selecting an efficient instruction ordering.

Figure 10-4 shows the dataflow graph of the code in Figure 10-1 (the interpretation of the numbers shown below each instruction will be discussed later). This dataflow graph is constructed by scanning the basic block, either from top to bottom or bottom to top, noting which instructions use data that is produced by other instructions in the basic block. An *edge* (arrow) in the dataflow graph signifies that data produced by one instruction is used by another instruction. Instructions that use data only from outside of the basic block are placed at the *root* (top) of the graph, and instructions producing data that is used only outside of the basic block are placed at the *leaves* (bottom) of the graph. There are more general methods for determining dependencies on data outside of the basic block, but these are not needed to schedule instructions within the basic block.

10.2.3 The Precedence Graph

The dataflow graph captures the *precedence constraints* that must be met by the instruction scheduler to satisfy true dependencies. That is, the dataflow graph specifies which instructions must precede other instructions so that data are produced before being used. However, the dataflow graph does not capture all precedence con-

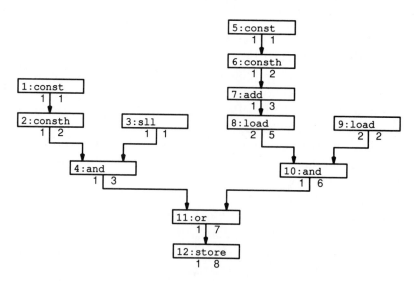

Figure 10-4. Dataflow Graph for Scheduling Example

straints. Some instructions must precede others because of antidependencies, output dependencies, ordering of memory references, or instruction side effects. These dependencies arise for the same reason they arise in hardware with out-of-order issue, the reason being that software might change the order of instructions just as hardware might. The fact that software changes the order statically rather than dynamically does not change the nature of the dependencies. A *precedence graph* captures these other precedence constraints in addition to the constraints due to true dependencies.

Figure 10-5 shows the additional edges (shown with dashed lines) that must be added to the dataflow graph of Figure 10-4 in order to show all precedence constraints. In this case, the additional edges between instructions 2, 4, and 5 are due to the anti- and output dependencies that arise from the reuse of the temporary register *tmp1*. The additional edges between instructions 8, 9, and 12 maintain the proper ordering of the loads with respect to the store, since there may be memory dependencies. The constraints between the loads and the store add no additional constraint than those due to true dependencies, and so do not limit the scheduler. However, the additional constraints between instructions 2, 4, and 5 force the scheduler to schedule instructions 2 and 4 before instruction 5 (assuming the code remains as given in

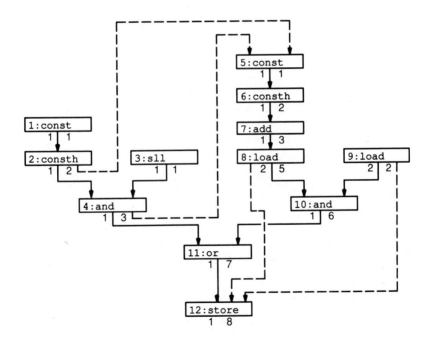

Figure 10-5. Precedence Constraints Caused by Temporary Register Reuse and Memory References

Figure 10-1). This is not desirable, since instruction 5 is otherwise independent, and could be scheduled to execute concurrently.

Hardware renaming does not help software to eliminate the precedence constraints due to anti- and output dependencies. Hardware renaming operates on the basis of the instruction ordering created by software and cannot help software reorder instructions in the presence of anti- and output dependencies.

However, the software scheduler is not necessarily constrained to accept a fixed allocation of registers. In fact, the schedule in Figure 10-2 assumes a different register allocation than in Figure 10-1 and has introduced a new temporary register *tmp3*. By reallocating registers, software can reduce the scheduling constraints of anti- and output dependencies that arise from register reuse, especially in the case of temporary registers. Reallocation is assumed in the remainder of this chapter, and the dataflow graph of Figure 10-4 is used to illustrate the scheduling process rather than the precedence graph of Figure 10-5. Even so, the software scheduler must in general be aware of precedence constraints that arise for reasons other than true dependencies.

10.2.4   The Concept of the Critical Path

In Figure 10-4, the instruction sequence 5, 6, 7, 8, 10, 11, and 12 constitute the *critical path* of this code fragment. The execution time of this path determines the minimum execution time of the entire sequence–no amount of instruction concurrency can make the execution time shorter. The numerical labels at the bottom of each operation in Figure 10-4 are used by the scheduler to determine the critical path of the sequence. The number to the left is the *operation latency*: that is, the latency from the beginning to the end of the operation (this is not necessarily the same as the result latency we have used in hardware, because an operation in a dataflow graph may include more than one instruction, as we will see later). The number to the right is the *path latency*: that is, the latency of the longest path from the top of the graph to the end of the operation. To compute the path latencies, the scheduler first assumes that the input operands to the first operations are available at the beginning of the sequence. Then, for each operation from the top to the bottom of the graph, the scheduler adds the operation latency to the path latency of the input operand(s) having the longest path latency. At the bottom of the graph, the latency of the critical path corresponds to the operation with the longest path latency. In Figure 10-4, the critical path is eight cycles long; note that the execution time of the scheduled instructions in Figure 10-2 is also eight cycles.

The critical path is important because it places a lower bound on the execution time of the scheduled instructions and identifies the operations that determine this execution time. The problem presented to the software scheduler is one of finding the shortest execution time that satisfies the constraints of processor resources and program correctness. In concept, this means that the scheduler is responsible for insuring that, where possible, operations on the critical path execute as quickly as the path length allows and that other operations execute in parallel with, and do not disrupt, the critical-path operations. As we will soon see, however, the scheduling problem is actually not quite so simple.

10.2.5 The Resource Reservation Table

The scheduler keeps track of resource usage and detects resource conflicts using a *resource reservation table*. The resource reservation table indicates the resources used by all operations, on a cycle-by-cycle basis. Figure 10-6 shows an example. When an operation is scheduled, the resources it requires are entered into the reservation table at the appropriate cycles. For example, Figure 10-6 shows entries for instructions 3 and 8 in the sequence of Figure 10-1; both instructions require result buses in the same cycle because of differences in latencies. When a new instruction is about to be scheduled, the scheduler checks the reservation table to determine whether the required resources are already being used. If there is a resource conflict, the scheduler can schedule the new instruction at a different time to avoid the conflict.

10.3 RELATIONSHIP OF THE SCHEDULER AND THE COMPILER

The scheduler and compiler together are responsible for translating a high-level-language program into code that can be efficiently executed by the superscalar processor. The final phases of the compiler–register allocation and code generation–operate in the same domain as the code scheduler, and there are several interactions between these compiler phases and the scheduler. This section describes these interactions and the problems and trade-offs they create.

10.3.1 Interaction of Register Allocation and Scheduling

All RISC processors have a large number of general-purpose registers, with the objective of providing enough registers that most program data can be supplied from high-speed, local storage. This requires that the compiler allocate registers to program variables to reduce the communication with the external data memory. In general, the goals of register allocation and of software scheduling are at odds with one

ALU1	ALU2	SHF	LS	BRN	R1	R2	Cycle
			8:load				n
		3:sll					n+1
					3	8	n+2

Figure 10-6. Resource Reservation Table

another. The register allocator wants to allocate as few registers to as much data as possible to decrease the possibility that there will not be enough registers. On the other hand, the scheduler wants to maintain as many independent computations as possible, meaning that additional registers are needed to store the intermediate results of parallel computations.

Section 10.2.3 pointed out that the reuse of registers by the register allocator can constrain the instruction schedule and reduce performance by giving rise to anti- and output dependencies in the code. Figure 10-7 illustrates the other side of this observation by showing the usage of registers in scheduled and unscheduled code. The usage of a register is the distance (here, in number of instructions) between the first instruction to define a value in the register and the last instruction to use a value in the same register (the value last used is not necessarily the value first defined, because the register may be written a number of times while it is being used). Scheduling has clearly increased register usage, even adding another register (*tmp3*). This increase

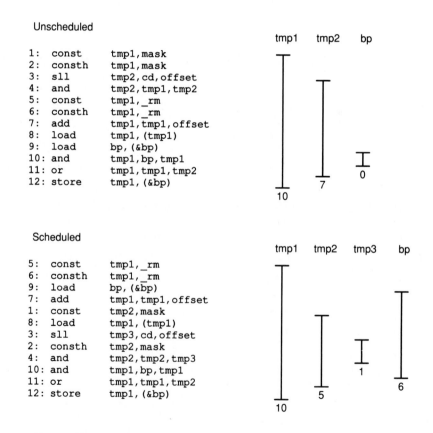

Figure 10-7. Register Usage in Unscheduled and Scheduled Code

in usage, taken along with the other values that the register allocator is attempting to keep in registers, reduces the chances that all required variables can be kept in registers.

The conflict between register allocation and scheduling presents a fundamental design trade-off. If register allocation is performed before scheduling, the scheduler may be unnecessarily constrained by the register allocation, particularly since the goal of the allocator is to reuse registers as much as possible. However, if scheduling is performed before register allocation, the register allocator may not be able to keep all variables in registers, because the scheduler increases register usage. When required data cannot be kept in registers, the time expense of external memory references can easily exceed the performance benefit of scheduling. In this chapter and Chapter 11, we dodge this issue by using the Am29000 architecture, which has such a large number of registers that the scheduler is not likely to be constrained by register allocation, and by using examples that do not require too many additional registers. This is not to say that scheduling is ineffective with fewer registers, or that it is a good idea to use registers indiscriminately in a processor with a large number of registers. Simply, the register architecture of the processor has an important bearing on the design of the scheduler with respect to register allocation.

Goodman and Hsu [1988] discuss an interesting solution to the conflict between register allocation and scheduling. They propose a scheduler that orders code for best performance when there are enough registers available and, when there are not enough registers, orders code for minimum register usage (which still obtains the best performance because it avoids the additional memory references caused by the inability to keep all operands in registers). This approach is particularly useful in a processor with fewer than 16 registers but also has some benefit in processors with 30-32 registers, such as the R2000.

10.3.2 Scheduling During Compilation Versus After Compilation

The software scheduler can be implemented in one of two ways. It can be part of the compiler, in which case compiling a program produces scheduled code. Alternatively, the scheduler can be independent of the compiler, in which case it operates on an assembly-language or object-code representation of a program. The latter alternative is sometimes called a *post-pass* scheduler [Hennessy and Gross 1983].

One of the primary advantages to scheduling in the compiler is that the compiler will likely have already determined basic blocks, dependency graphs, and other information required by the scheduler. A post-pass scheduler must recompute this information. The scheduler in the compiler also has access to the compiler's register allocator, so that register allocation can be determined, at least in part, by the needs of the scheduler. A post-pass scheduler could reallocate registers to reduce storage dependencies and create code whose quality is comparable to that of a scheduling compiler, but this is usually considered too difficult. The post-pass scheduler just schedules within the constraints of the storage dependencies existing in the code.

In addition, the compiler is likely to have detailed information that does not exist in the object code. For example, a variable access in the compiler appears as a load or store in the final code. A post-pass scheduler usually cannot determine which variable is being accessed by a load or store and cannot determine the dependencies between loads and stores. Knowing what variables are being accessed allows the scheduler to reorder memory references and improve parallelism when there are no dependencies.

Another minor advantage is that, when the scheduler is implemented in the compiler, the scheduler may help guide instruction selection. When there is more than one choice of instruction sequences for a sequence of operations, the scheduler can help make the best choice. Also, if the compiler provides information for an interactive debugger, the scheduler is less likely to invalidate this information if scheduling is performed by the compiler.

Not all programming languages allow code to be arbitrarily intermingled. The version of the C programming language specified by the 1988 draft of the American National Standards Institute (ANSI) has this sort of restriction [ASC/X3 1988]. The ANSI C specification dictates that semicolons in the source program (which end source statements) are *sequence points*. The instructions that implement the language statements are restricted in ability to move across the sequence points. The reason for this restriction is that, if the program is interrupted, it is desirable to associate the processor and system state with a point in the original source program. Restrictions such as these mandate that the scheduler be incorporated into the compiler, because only the compiler knows about the sequence points in the source program.

Precedence rules for expressions may also restrict the code motions that the scheduler can perform. For example, consider the following source statement:

$$X = ((A+B)+C)+D$$

Without the precedence constraints imposed by the parentheses, the scheduler could reorder this computation as

$$X = (A+B)+(C+D)$$

so that the additions $A+B$ and $C+D$ can be performed at the same time. However, the original statement prohibits this reordering. The placement of parentheses in the original statement do not allow the sum of C and D to appear as an intermediate term: C must be first added to the sum $A+B$, and then D added to the result. If the scheduler does not adhere to this order, the computed value may be incorrect. For example, if these are floating-point computations, the intermediate values computed by the two different orderings may be rounded differently and generate different final results. Only the compiler knows the precedence rules for the language and knows which operations are and are not subject to precedence constraints. Only a scheduler in the compiler could see the appropriate set of precedence constraints and work in concert with code generation to emit the appropriate sequence of instructions. A post-pass scheduler could not improve the performance of such a sequence, because it cannot know the precedence constraints and cannot affect which instructions appear in the code.

A minor criticism of scheduling in the compiler is that there is a circularity between scheduling and code generation. Some instructions selected by the code generator may have specific instruction interactions that interfere with scheduling, such as the setting and use of condition codes, whereas other, equivalent instruction sequences might not impose these constraints. Furthermore, if the scheduler operates on intermediate code representations, scheduling decisions can be undone because the operation ordering chosen by the scheduler has no direct representation as an instruction sequence. These interdependencies make it difficult to realize the full benefit of cooperation between the code generator and the scheduler without an inordinate amount of complexity, because both the scheduler and the code generator have to be aware of the constraints imposed on the other. These problems particularly apply to CISC processors, which can have complex interactions between instructions and which have less correspondence between the compiler's intermediate code format and the final generated code. In practice, this just means that the full theoretical benefit of scheduling in the compiler is not realized.

10.4 ALGORITHMS FOR SCHEDULING BASIC BLOCKS

The task of a software scheduler is simply to select the order of a sequence of instructions so that they execute correctly in a minimum amount of time, using the given processor resources. However, a software scheduler requires an enormous amount of time to guarantee that a schedule is optimum: that is, to guarantee that the schedule has a minimum execution time. Such a guarantee requires that the scheduler exhaustively test the execution time of every possible correct schedule. Instead of taking this time-consuming approach, most schedulers sacrifice the guarantee of optimality but often generate a schedule that is optimum or nearly optimum anyway. This relies on the use of *heuristics*, or algorithms that seem reasonable and usually give good results even though they do not guarantee good results.

10.4.1 The Expense of an Optimum Schedule

Software scheduling is easy for an ideal processor with infinite resources, because the execution time of the code is determined only by data dependencies. Because the processor has infinite resources, there are always available resources to execute operations off of the critical path in parallel with the critical path, and these operations cannot create resource conflicts with the critical path. If there can be no resource conflicts, the scheduler simply determines the critical path from the data dependencies in the code and schedules operations so that the critical path is processed in the minimum possible time. This schedule is guaranteed to be optimum–in effect, the dataflow graph and the optimum schedule are the same thing.

Scheduling is significantly more complex in a real processor with finite resources, because the execution time of the code is determined both by data dependencies and by resource conflicts. In general, the execution time is determined not only by the critical path but also potentially by other paths, depending on the allocation of resources to operations and on the data dependencies between these operations. At the point of scheduling any given operation, the scheduler cannot be sure that the code order chosen for the operation will not create a resource conflict with another operation, with the result that the overall execution time is less than optimum.

The scheduler can guarantee that a schedule is optimum by trying all possible schedules that satisfy the precedence constraints and choosing the schedule that has the minimum execution time. However, this approach causes explosive growth in the time that the scheduler takes to compute the schedule. For example, in the dataflow graph of Figure 10-4, there are 4 choices for the first instruction, 14 choices for the ordering of the first two instructions, and 40 choices for the ordering of the first three instructions. Even though the scheduler can quickly determine that many of these alternatives yield no improvement in execution time, the sheer number of basic blocks in an average program and the number of alternatives in each block causes an unacceptably long execution time for the scheduler. For example, Davidson and colleagues [1981] reported that an optimum scheduler for small programs required three to four times the scheduling time of heuristic schedulers. For one large program, the scheduling time exceeded the maximum execution time allowed by the Multics operating system and could not be completed. The estimated running time was on the order of weeks or years–obviously impractical.

The execution time taken to guarantee an optimum schedule is particularly unacceptable for programs that do not have much instruction parallelism. If only a few instructions can be issued per cycle, the processor can easily have a larger number of functional units than are required for these few instructions, on the average. In this case, the processor is closer to having an "infinite" number of resources, from the scheduler's point of view, compared to a processor that has a number of functional units that is about the same as the number of instructions that can be issued. With small amounts of parallelism, the execution time is more likely to be constrained by data dependencies than by resource conflicts, so a schedule that is determined mainly by the precedence graph is likely to be optimum.

10.4.2 List Scheduling

There are a number of software scheduling algorithms. Each algorithm makes a trade-off between the time taken by the scheduler and the presumed optimality of the resulting schedule; none, however, can guarantee an optimum schedule without excessive computation time.

Rather than describing all the alternatives, this section illustrates scheduling principles using *list scheduling*. List scheduling encompasses a class of algorithms that schedule operations one at a time from a list of operations to be scheduled, using prioritization to resolve contention for functional units. List scheduling produces optimum results in almost all practical cases [Adam et al. 1974, Fisher 1979] and is fast because it does not backtrack to reconsider earlier scheduling decisions. This section describes a popular version of list scheduling and the rationale behind it. The following sections explore some of the finer points of this algorithm and briefly introduce some alternative algorithms.

The list-scheduling algorithm described here schedules operations from the bottom to the top of the precedence graph, in reverse order to their actual execution and going backward in time (Section 10.4.3 explains the reason for this). Operations are scheduled by building a sequence of instructions, adding each new instruction to the top of the previously scheduled sequence, starting with the last instructions in the sequence. As operations are scheduled, they make preceding operations available for scheduling. If there is a conflict between operations for a processor resource, this conflict is resolved in favor of the operation on the longest dependency path. This is accomplished by assigning to each operation a priority equal to the path latency of its outputs, relative to the top of the dataflow graph (see Figure 10-4). The operations with longer path latencies are given higher priority during scheduling, because these operations are assumed to have more effect on the overall execution time of the sequence than operations with shorter latencies.

Before scheduling begins, the operations at the bottom of the dataflow graph are added to the set of operations that are ready to be scheduled (the *ready set*). A cycle counter is set at the cycle in which these operations are to be scheduled (conceptually, this is the last cycle in the sequence, and the actual value does not matter). Then, as long as there are operations still to be scheduled, the following steps are performed:

1. The operation in the ready set with the highest priority is scheduled, unless it requires a resource that is busy in this cycle. If all required resources are available, the instruction is added to the top of the currently scheduled sequence.

2. If the operation is scheduled in step 1, the resources used by the operation are set busy for the appropriate cycles in the resource reservation table (if this operation causes the resource to be busy for more than one cycle, such as when a functional unit is not sufficiently pipelined to be able to accept an instruction on

---

every cycle, then the resource is also set busy for the appropriate number of previous cycles).

3. If the operation is scheduled in step 1, it is removed from the ready set. If the operation is not scheduled because of a resource conflict, it remains in the ready set and scheduling continues with other operations in the set.

4. Steps 1 through 3 are repeated until scheduling has been attempted for all operations in the ready set.

5. The cycle counter is decremented to the previous cycle.

6. Each operation not in the ready set whose successors have all been scheduled is examined to determine whether or not it is ready for scheduling. An operation is ready when the time interval between the current cycle and the scheduled cycle of the nearest successor is equal to the latency of the operation in question (the nearest successor is the one with the earliest scheduled execution time). This reflects the fact that the operation in question must execute soon enough to satisfy the dependencies of its immediate successors (recall that scheduling, conceptually, proceeds backward in time). For example, an operation with a three-cycle latency is added to the ready set three cycles before its nearest successor. Operations that become ready are added to the ready set.

7. If there are still operations to be scheduled, go to step 1. Otherwise, scheduling is complete.

Table 10-1 shows the cycles in which the operations of Figure 10-4 are added to the ready set, along with the priorities of each operation. Figure 10-8 shows each step of the scheduling procedure, and Figure 10-2 shows the final schedule. Note in particular steps 8, 9, and 10. The *load* operation 9 is not scheduled in cycle 4 because it conflicts with the *load* operation 8 that has higher priority. Instead, it is scheduled

Table 10-1. Scheduling Order and Priority Assignment for Operations in Dataflow Graph of Figure 10-4

Cycle	Nodes/Priorities
8	12/8
7	11/7
6	10/6, 4/3
5	2/2, 3/1
4	8/5, 9/2, 1/1
3	7/3
2	6/2
1	5/1

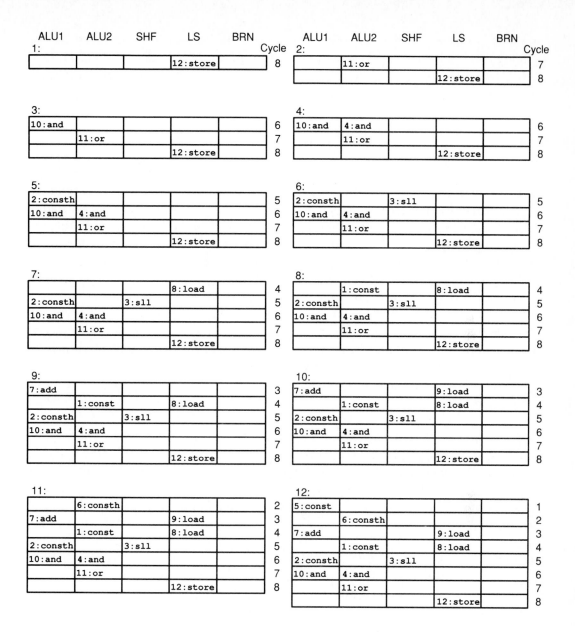

Figure 10-8. List-Scheduling Procedure

in cycle 3 after the *add* operation 7 (which also has higher priority, though there are no resource conflicts in this case).

The resource table in Figure 10-8 shows only the functional units for clarity, but other processor resources, such as result buses, can also have table entries and affect

---

the scheduling process. In this example, conflicts for result buses are ignored, assuming that hardware arbitration will resolve conflicts. In practice, this can degrade performance, because the scheduler does not consider all resource conflicts.

10.4.3 The Effect of Scheduling Order

Scheduling operations from the bottom to the top of the precedence graph tends to schedule operations just-in-time: that is, an operation is scheduled in the last possible cycle that does not increase the overall execution time of the sequence. This occurs because operations are released for scheduling by their successors being scheduled, rather than by their input operands being ready. Scheduling from top to bottom would have the opposite effect, in that operations would be scheduled as soon as their input operands were ready, regardless of when the output results were needed. Another way to conceptualize this is to consider that there are several paths through the code that are shorter than the critical path. Scheduling from bottom to top tends to place these shorter paths toward the end of the sequence so that they execute with the end of the critical path. Scheduling from top to bottom tends to place these operations toward the beginning of the sequence so that they execute with the beginning of the critical path.

Scheduling from top to bottom has the benefit that results are generally produced as soon as possible, given the available resources. Figure 10-9 shows how scheduling from top to bottom tends to schedule operations in earlier cycles (understand that the priorities used to schedule here are the path latencies relative to the bottom of the precedence graph rather than the top). Producing results early has two benefits. First, a basic block can contain operations that do not appear to be on the critical path of the current basic block but that are actually on a global critical path that spans two or more basic blocks. Scheduling operations early incurs less risk that the global critical path is lengthened, because most operations are started as soon as the processor resources allow. Second, scheduling operations early tends to move the evaluation of branch conditions up in the schedule, reducing the average time taken to detect a misprediction and increasing instruction throughput. The scheduler is not able to move a branch from the bottom of a basic block, and, because of this, scheduling from bottom to top tends to place the evaluation of branch conditions at the bottom of the basic block, just before the branch. This increases the average time taken to detect a branch misprediction.

Figure 10-10 shows how the execution of the bottom-to-top schedule compares to the execution of the top-to-bottom schedule, illustrating that instructions are executed earlier and that results are generally available earlier. Note that the availability of results in the top-to-bottom schedule is limited by conflicts for the result buses.

ALU1	ALU2	SHF	LS	BRN
5:const	1:const	3:sll	9:load	
6:consth	2:consth			
7:add	4:and			
			8:load	
10:and				
	11:or			
			12:store	

```
 5:  const    tmp1,_rm           ;base address rm
 1:  const    tmp2,mask          ;load mask
 9:  load     bp,(&bp)           ;*bp
 3:  sll      tmp3,cd,offset     ;cd<<offset
 6:  consth   tmp1,_rm
 2:  consth   tmp2,mask
 7:  add      tmp1,tmp1,offset   ;address of rm[offset]
 4:  and      tmp2,tmp2,tmp3     ;(cd<<offset) & mask
 8:  load     tmp1,(tmp1)        ;rm[offset]
10:  and      tmp1,bp,tmp1       ;*bp & rm[offset]
11:  or       tmp1,tmp1,tmp2     ;final expression
12:  store    tmp1,(&bp)         ;assign *bp
```

Figure 10-9. Schedule and Code Obtained by List Scheduling from Top to Bottom of the Precedence Graph

10.4.4 Other Scheduling Alternatives

We have examined list scheduling in detail, but there are many other scheduling algorithms. Scheduling algorithms differ in the ways that they consider the precedence graph, in the order that they select operations for scheduling, in the number of operations they consider at once, and in the criteria they use to resolve conflicts for functional units. This section briefly considers the ideas behind some of the alternative approaches to scheduling. Davidson and colleagues [1981] provide a good overview of different alternatives and provide references to other publications.

A *first-come, first-served* scheduler scans through a basic block in some order (such as the instruction order in an object file). Each instruction is moved as far up in the block as its dependencies will allow. This simple, fast approach tends to schedule operations as early as possible. However, it considers instructions in an arbitrary order and ignores the critical path.

A *greedy* scheduler computes the schedule for all instructions along one given path in the precedence graph before scheduling other paths. The critical path is scheduled first, the next shortest path second, and so on. This gives highest priority to the critical path, but may miss cases where the execution time is not actually deter-

Bottom to Top

I0	I1	I2	I3
5:const	6:consth	9:load	7:add
1:const	8:load	3:sll	2:consth
4:and	10:and	11:or	12:store

Decode

ALU1	ALU2	SHF	LS	BRN	R1	R2	Cycle
							1
5:const			9:load				2
	6:consth	3:sll			5		3
7:add	1:const		——		6	3	4
2:consth	4:and		8:load		7	1	5
	——				9	2	6
10:and					4	8	7
	11:or				10		8
			12:store		11		9

Execute / *Writeback*

Top to Bottom

I0	I1	I2	I3
5:const	1:const	9:load	3:sll
6:consth	2:consth	7:add	4:and
8:load	10:and	11:or	12:store

Decode

ALU1	ALU2	SHF	LS	BRN	R1	R2	Cycle
							1
5:const	1:const	3:sll	9:load				2
6:consth	2:consth	——			5	1	3
7:add	——		——		3	6	4
	4:and		8:load		2	7	5
					9	4	6
10:and					8		7
	11:or				10		8
			12:store		11		9

Execute / *Writeback*

Figure 10-10. Result Availability: Scheduling Bottom to Top Versus Top to Bottom

mined by the longest dependency path, but by some other path because of resource conflicts.

A *branch-and-bound* scheduler considers all possible instruction orderings and selects the optimum ordering. Since the schedule is optimum, this approach potentially takes a very long scheduling time. In order to limit the scheduling time, the scheduler discards an alternative as soon as it discovers that the alternative is not faster than the current best alternative it has found. Even with this optimization, a branch-and-bound scheduler takes too long to schedule if there are a large number of instructions. However, this algorithm may be acceptable for small sequences of code or in particular situations.

Scheduling algorithms can have varying degrees of *lookahead,* meaning that they consider some number of future scheduling options opened up by a given scheduling decision. Other algorithms achieve the same effect by *backtracking* to reconsider earlier scheduling decisions. Considering more alternatives during scheduling presumably leads to a better schedule, though it also leads to longer scheduling time. Complex scheduling algorithms also have the possibility of *deadlock*, due to the interaction between the criteria used to select operations for scheduling and the criteria used to actually schedule the operations. It is possible for the scheduler to get into a state in which it will not examine any node that it will also choose to schedule, in which case the scheduler is faulty.

The most important criteria for choosing a scheduling algorithm is that the algorithm generates an optimum or nearly optimum schedule for a large number of practical applications and that the scheduler does not take forever to generate the schedule. Generating a good schedule usually involves considering the critical path, and having an acceptable scheduling time usually means not being too fancy (for example, having little or no backtracking).

10.5 REVISITING THE HARDWARE

Software instruction scheduling is important for a superscalar processor, because it exposes instruction parallelism to the hardware. The potential benefit of scheduling in a superscalar processor is greater than the benefit in a scalar processor, because there are more opportunities to improve performance and because each cycle saved is a larger proportion of the overall execution time.

But what effect does software scheduling have on the design of the hardware resources? Software scheduling increases hardware utilization and the peak hardware demand. For example, scheduled code generates the same number of results as the unscheduled code in a shorter amount of time, and thus creates more contention for the result buses. It is hard to say whether the increased demand for functional units and result buses is severe enough to justify adding more of them, particularly because of the cost of these resources.

Scheduling also has some effect on the sizes of the instruction window and the reorder buffer. We can get an idea of these effects from Figure 10-11, which shows the utilization of the instruction window and the reorder buffer (in terms of the number of entries used) for both scheduled and unscheduled code. Scheduling reduces the average utilization of the instruction window: the unscheduled code uses an average of 4.6 entries per cycle and the scheduled code uses an average of 4.1 entries per cycle. Scheduling has the opposite effect on the utilization of reorder buffer entries: the unscheduled code uses an average of 7 entries per cycle and the scheduled code uses an average of 9.25 entries per cycle. This single example cannot be used to gauge the magnitude of the overall effect that scheduling has on the utilization of the instruction window or the reorder buffer, but the trend generally holds. Scheduling reduces the utilization of the instruction window because an instruction placed into

```
*bp = (*bp & rm[offset]) | ((cd<<offset) & mask)
```

Unscheduled

	Fetch/Decode			Execute					Window	Reorder buffer	Cycle
I0	I1	I2	I3	ALU1	ALU2	SHF	LS	BRN			
1:const	2:consth	3:sll	4:and						–	–	1
5:const	6:consth	7:add	8:load	1:const		3:sll			4	4	2
9:load	10:and	11:or	12:store		2:consth				6	8	3
				4:and	5:const				9	12	4
				6:consth					7	12	5
					7:add				6	10	6
							8:load		5	8	7
							9:load		4	6	8
									4	5	9
				10:and					3	5	10
					11:or				2	4	11
							12:store		1	3	12

avg 4.6 and 7/cycle

Scheduled

	Fetch/Decode			Execute					Window	Reorder buffer	Cycle
I0	I1	I2	I3	ALU1	ALU2	SHF	LS	BRN			
5:const	6:consth	9:load	7:add						–	–	1
1:const	8:load	3:sll	2:consth	5:const			9:load		4	4	2
4:and	10:and	11:or	12:store		6:consth	3:sll			6	8	3
				7:add	1:const	——			8	12	4
				2:consth	4:and		8:load		6	12	5
					——				3	10	6
				10:and					3	10	7
					11:or				2	10	8
							12:store		1	8	9

avg 4.1 and 9.25/cycle

Figure 10-11. Window and Reorder-Buffer Utilization of Unscheduled and Scheduled Code

the window is more likely to be ready to issue than with unscheduled code. Conversely, scheduling increases the utilization of the reorder buffer, because more instructions are completed in parallel, and because the execution time is shorter.

However, software scheduling raises a question that is more important than these minor hardware adjustments. Figure 10-12 shows the execution timing of the scheduled code on a processor with in-order issue, rather than out-of-order issue. The execution time of this sequence is exactly the same as the execution time with out-of-order issue, and both execution times are better than the execution time of un-

Decode				Execute					Writeback		Cycle
I0	I1	I2	I3	ALU1	ALU2	SHF	LS	BRN	R1	R2	
5:const	6:consth	9:load	7:add								1
	6:consth	9:load	7:add	5:const							2
			7:add		6:consth		9:load		5		3
1:const	8:load	3:sll	2:consth	7:add					6		4
			2:consth		1:const	3:sll	8:load		7	9	5
4:and	10:and	11:or	12:store	2:consth					1	3	6
		11:or	12:store	10:and	4:and				2	8	7
			12:store		11:or				10	4	8
							12:store		11		9

Figure 10-12. Execution Timing of Scheduled Code with In-Order Issue

scheduled code with out-of-order issue. Obviously, in-order issue with software scheduling is better than out-of-order issue without software scheduling, at least within the confines of a basic block. Out-of-order issue does not improve the execution time, whereas software scheduling does. Software is much better than hardware at uncovering parallelism within a basic block, because software is able to consider many different possible orderings of instructions in the basic block. Hardware is not able to improve on the software schedule; hardware examines only a few instructions at once and is not nearly as capable as software. Furthermore, software scheduling can perform other hardware functions that we have examined, such as insuring that register dependencies are enforced. We have seen that hardware register renaming is not of much use in helping the software scheduler to uncover parallelism between instructions. Within the bounds of a basic block, there is little justification for the complex hardware we have studied.

The best we can say for out-of-order issue at this point is that it does not use the decoder as much as in-order issue. With in-order issue (Figure 10-12), the decoder throughput is 1.5 instructions per cycle. With out-of-order issue (Figure 10-11), the decoder throughput is 4 instructions per cycle. With out-of-order issue, the decoder is available to accept more instructions at the end of 3 cycles. With in-order issue, the decoder is not available until after 8 cycles.

In general, counting alignment penalties and assuming branch prediction is correct, the average decoder throughput with out-of-order issue is about two instructions per cycle. This throughput is achieved by fetching and executing instructions from adjacent basic blocks. For this reason, out-of-order issue may still have higher performance than in-order issue with software scheduling–possibly as much as 30% higher in this example if we assume that the decoder throughput of 1.5 instructions per cycle can be raised to an average of 2 instructions per cycle with out-of-order issue. Software would have to move 4 instructions into this code sequence from another basic block to achieve the same average decoder throughput and the same performance advantage. We cannot resolve the question of software scheduling versus out-of-order issue until we examine software techniques for scheduling across branches. This is the topic of the next chapter.

Chapter 11

Software Scheduling Across Branches

We can simplify the superscalar processor by eliminating out-of-order issue, using instead in-order issue with software scheduling to improve performance. This is the principal reason for the material presented in the preceding chapter: we are looking for ways to simplify the superscalar processor. However, software scheduling, when confined within the bounds of the basic block, does not yield nearly the same benefit to overall performance as does out-of-order issue. This is true though we just concluded in the previous chapter that, within the basic block, software scheduling is superior to out-of-order issue. The reason for this apparent contradiction is that out-of-order issue is not confined to the basic block, and it allows the hardware to look outside of the basic block for parallel instructions.

Figure 3-10 (page 54) gives us an idea of the best performance we should expect from in-order issue with basic-block scheduling by software. In Figure 3-10, the average speedup with in-order issue and out-of-order completion is 1.22. This speedup is achieved with hardware that decodes an entire basic block of instructions in one cycle. Since the processor also renames registers, it can issue instructions in the basic block in any order permitted by true dependencies, without concern for storage conflicts. The only restriction on instruction issue is that an entire basic block must be in execution before the next block can begin execution. This approximates the best performance achieved by basic-block scheduling with in-order issue, missing only the cases where software scheduling might avoid resource conflicts, but being optimistic about the processor's ability to fetch instructions. But performance is far below the speedup of 1.94 achieved by out-of-order issue. Scheduling instructions within a basic block, as presented in the previous chapter, has a limited ability to improve performance, because it cannot move instructions across basic-block boundaries. Though scheduling basic blocks causes the execution time of each basic block to be nearly optimum, this does not necessarily cause the execution time of the overall program to be nearly optimum, because the processor hesitates at each branch until all instructions preceding and including the branch are in execution.

The ability to move instructions across basic-block boundaries gives the software scheduler more flexibility to create a good schedule and provides a good source of parallel operations. Also, if the software scheduler knows about critical paths that span several basic blocks, it may make better scheduling decisions than if it knows only about the critical path within each basic block. This ability is potentially attractive in general-purpose applications, because, as we have seen, these applications tend to have a large number of branches and small basic blocks having little parallelism within the block.

Despite these advantages, the scheduler is not permitted to move instructions arbitrarily across branch boundaries, because the resulting code is probably incorrect. Branches determine which instructions should be executed, or, more precisely, what items of processor and system state the program should modify. The scheduled code should make all of the modifications that are required and should make no modifications other than these. This obviously restricts instruction movement with respect to branches. Furthermore, there are many possible critical paths because there are many possible paths through the code. When scheduling across basic-block boundaries, the scheduler must make assumptions about the likely critical paths. This chapter considers scheduling techniques that can move instructions across branches but still maintain the correctness of the code. As we will see, scheduling across branches often relies on a notion of "correctness" that may or may not be acceptable, depending on the system.

The scheduling techniques presented in this chapter are illustrated with a processor that issues instructions in order, to show more clearly the benefit of scheduling across branches. A processor that predicts branches and issues instructions out of order can also essentially schedule instructions across branches, independent of software scheduling. Showing the behavior of scheduled code on hardware with out-of-order issue tends to obscure the benefit of software scheduling for the simple examples presented here. This is more than a simple concern over presenting material in a clear fashion: it is an indication that software scheduling across branches and out-of-order issue have very similar effects on program performance.

For clarity, delayed branches are not used in the examples presented in this chapter, though most RISC processors (including the Am29000 used in the examples) have delayed branches.

11.1 TRACE SCHEDULING

Trace scheduling [Fisher 1979, 1981; Ellis 1986] originally was developed as a technique for scheduling and packing operations into *horizontal microinstructions* (a horizontal microinstruction is an instruction that directly controls hardware and that specifies many operations). Trace scheduling has evolved into the principal technique used in very-long-instruction-word (VLIW) architectures for scientific applications [Fisher 1983, Colwell et al. 1987]. This section shows how trace scheduling can be adapted to a superscalar architecture.

A *trace* is a possible path through a section of code, usually taken to mean a path that spans more than one basic block. The sequence of instructions in the trace is determined by assuming a particular outcome for every branch in the sequence. There is a separate trace for each combination of branch outcomes. For example, in Figure 11-1 there are two traces, depending on the outcome of the *if* statement. (This is nearly the same meaning of "trace" as it applies to our simulator, because the simulator analyzes a dynamic path through a large section of a program. The only difference is that trace scheduling deals with a static trace.)

11.1.1 A Simple Example of Trace Scheduling

Trace scheduling is similar to basic-block scheduling, except that entire traces are scheduled rather than single basic blocks. Figure 11-2 shows a code segment that we will use to illustrate trace scheduling. There are two basic blocks in this example; both basic blocks contain instructions that have little parallelism because of true dependencies. Figure 11-2 also shows the execution of these basic blocks by a processor with in-order issue. The total execution time is 12 cycles.

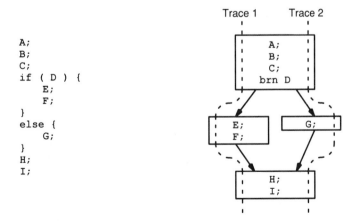

Figure 11-1. Examples of Traces in Code

```
a[i] = a[i] + 1;
if ( a[i] < 100 ) {
    count = count + 1;
    b->sum = b->sum + a[i];
}
```

```
 1:  sll      tmp1,i,2             ;scale index
 2:  add      tmp1,tmp1,base_a     ;add to base of a
 3:  load     tmp2,(tmp1)          ;a[i]
 4:  add      tmp2,tmp2,1          ;a[i]+1
 5:  store    tmp2,(tmp1)          ;assign a[i]
 6:  cplt     tmp1,tmp2,100        ;a[i]<100?
 7:  jmpf     tmp1,LABEL           ;if branch  .
 8:  add      count,count,1        ;count=count+1
 9:  add      tmp1,base_b,s_off    ;address of b->sum
 10: load     tmp3,(tmp1)          ;b->sum
 11: add      tmp3,tmp3,tmp2       ;b->sum+a[i]
 12: store    tmp3,(tmp1)          ;assign b->sum
LABEL: 13: (statements following if)
```

| Decode | | | | Execute | | | | | Writeback | | |
I0	I1	I2	I3	ALU1	ALU2	SHF	LS	BRN	R1	R2	Cycle
1:sll	2:add	3:load	4:add								1
	2:add	3:load	4:add			1:sll					2
		3:load	4:add	2:add					1		3
			4:add				3:load		2		4
			4:add								5
5:store	6:cplt	7:jmpf	8:add		4:add				3		6
		7:jmpf	8:add	6:cplt			5:store		4		7
9:add	10:load	11:add	12:store		8:add			7:jmpf	6		8
	10:load	11:add	12:store	9:add					8		9
		11:add	12:store				10:load		9		10
		11:add	12:store								11
			12:store		11:add				10		12
							12:store		11		13

Figure 11-2. Code for Trace Scheduling and Execution Timing with In-Order Issue

The effectiveness of trace scheduling depends on software knowing the likely execution traces, because, as we will see, the execution time of some traces is reduced at the expense of other traces. It is important that trace scheduling reduce the execution time of likely traces at the expense of unlikely ones, instead of the other way around. To predict the likely execution traces, software must have a way to predict the likely outcomes of branches, using profiles of previous program executions or reasonable guesses. For the purpose of illustrating trace scheduling on the code of Figure 11-2, the branch of the *if* statement is assumed not taken. The two basic blocks are merged into a single sequence of instructions, and the resulting sequence is scheduled using the list-scheduling algorithm described in Section 10.4.2. During scheduling, this sequence is treated as a single basic block, and the branch is treated as any other instruction.

The code resulting from trace scheduling is shown in Figure 11-3. Figure 11-3 also shows the execution sequence of this code, again on a processor with in-order issue. Trace scheduling has overlapped the serial computations in the two basic blocks, reducing the execution time from 12 cycles to 8. The load of *b–>sum* (instruction 10) has been moved before the assignment to *a[i]* (instruction 5), under the assumption that these values do not occupy the same memory location and therefore cannot be dependent. However, this code as it stands has other problems, because it does not account for cases where the *if* branch is taken.

```
a[i] = a[i] + 1;
if ( a[i] < 100 ) {
        count = count + 1;
        b->sum = b->sum + a[i];
}
```

```
        1:  sll      tmp1,i,2             ;scale index
        9:  add      tmp2,base_b,s_off    ;address of b->sum
       10:  load     tmp3,(tmp2)          ;b->sum
        2:  add      tmp1,tmp1,base_a     ;add index to base of a
        3:  load     tmp4,(tmp1)          ;a[i]
        4:  add      tmp4,tmp4,1          ;a[i]+1
       11:  add      tmp3,tmp3,tmp4       ;b->sum+a[i]
        6:  cplt     tmp5,tmp4,100        ;a[i]<100?
        5:  store    tmp4,(tmp1)          ;assign a[i]
        8:  add      count,count,1        ;count=count+1
       12:  store    tmp3,(tmp2)          ;assign b->sum
        7:  jmpf     tmp5,LABEL           ;if branch
LABEL: 13:  (statements following if)
```

| | Decode | | | | Execute | | | | Writeback | | |
I0	I1	I2	I3	ALU1	ALU2	SHF	LS	BRN	R1	R2	Cycle
1:sll	9:add	2:add	10:load								1
		2:add	10:load		9:add	1:sll					2
3:load	4:add	11:add	6:cplt	2:add			10:load		1	9	3
	4:add	11:add	6:cplt				3:load		2		4
	4:add	11:add	6:cplt						10		5
			6:cplt	11:add	4:add				3		6
5:store	8:add	12:store	7:jmpf		6:cplt				11	4	7
		12:store	7:jmpf	8:add			5:store		6		8
							12:store	7:jmpf	8		9

Figure 11-3. Trace Schedule for Code of Figure 11-2 (Incorrect)

11.1.2 Using Compensation Code to Recover from Incorrect Predictions

The code in Figure 11-3 assumes that the *if* branch is not taken. When this branch is taken, instructions from the second basic block have been executed when they should not have been, because trace scheduling moved them to the first basic block. To correct this problem, while retaining the advantage of trace scheduling, *compensation code* is added to provide correct results in case the branch is taken.

Compensation code normally appears *off the trace,* in basic blocks that are not part of the scheduled trace but that are executed when the outcomes of branches in the trace are different than the predicted outcomes, or in basic blocks that branch into the middle of the trace. Because trace scheduling can move an instruction up or down in the trace and can move an instruction through either a branch or the target of a branch, there are four types of compensation code.

When an instruction in the trace is moved down after the branch at the end of its basic block, that instruction must be copied to the block that is executed when the branch has the alternate outcome (refer to the movement of instruction *C* in Figure 11-4a). This insures that the instruction is executed when the branch in the trace has the unexpected outcome. When an instruction is moved down to a basic block that can be reached via paths off the trace, the branches in these other paths are adjusted so that they join the trace beyond the moved instructions (refer to the movement of instruction *F* in Figure 11-4a). Moving the point where other paths join the trace may cause these paths to skip instructions that should be executed, and the

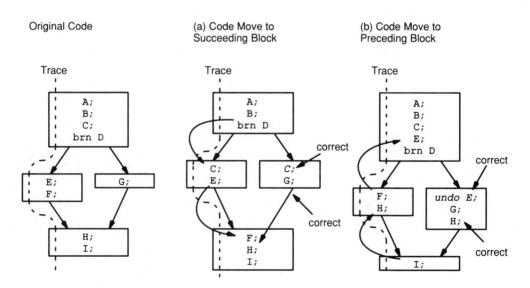

Figure 11-4. Compensation Code Added During Trace Scheduling

skipped instructions must be copied to the paths off the trace to avoid this. If the instruction is moved through more than one succeeding branch, compensation must be repeated at each basic block.

When an instruction is moved up in the trace from a block that can be reached via paths off the trace, that instruction must be copied to all blocks that are off the trace (refer to the movement of instruction *H* in Figure 11-4b). This insures that the instruction is executed regardless of the path taken through the code. When an instruction is moved up before the branch in a preceding basic block, any changes to permanent state (for example, program variables) must be undone in the block that is executed when the branch has the alternate outcome (refer to the movement of instruction *E* in Figure 11-4b). If the instruction is moved up through more than one basic block, compensation must be repeated at each basic block.

Figure 11-5 shows the scheduled code of Figure 11-3 with compensation code added on the lines labeled *c1, c2,* and *c3.* In this example, all instructions in the second basic block have been moved to the preceding basic block, so the effect of these instructions must be undone when the *if* branch is taken. The scheduler has had to create a third basic block for the compensation code (starting at the label *COR-RECT*), because, in the original code, the basic block following the *if* statement is executed regardless of the outcome of the *if.* The compensation code should be executed only when the *if* branch is taken. This results in a messy branch sequence, but this can be corrected in a later step by complementing the test of the *if* branch and removing the jump around the correction code.

As the foregoing discussion illustrates, the compensation code is ad hoc and very dependent on the instruction set of the processor. Sometimes the effects of an instruction can be undone by an inverse operation. For example, in Figure 11-5 the

```
          1: sll      tmp1,i,2            ;scale index
          9: add      tmp2,base_b,s_off   ;address of b->sum
          10: load    tmp3,(tmp2)         ;b->sum
          2: add      tmp1,tmp1,base_a    ;add index to base of a
          3: load     tmp4,(tmp1)         ;a[i]
          4: add      tmp4,tmp4,1         ;a[i]+1
          11: add     tmp5,tmp3,tmp4      ;b->sum+a[i]
          6: cplt     tmp6,tmp4,100       ;a[i]<100?
          5: store    tmp4,(tmp1)         ;assign a[i]
          8: add      count,count,1       ;count=count+1
          12: store   tmp5,(tmp2)         ;assign b->sum
          7: jmpf     tmp6,CORRECT        ;if branch
          c1: jmp     LABEL               ;empty block
 CORRECT: c2: sub     count,count,1       ;(correct count)
          c3: store   tmp3,(tmp2)         ;(correct b->sum)
   LABEL: 13: (statements following if)
```

Figure 11-5. Trace Schedule for Code of Figure 11-2 with Compensation Code

statement *count = count + 1* is undone by the operation *count = count − 1.* This is rarely, if ever, done. In general, instead of relying on an inverse operation, the scheduler introduces instructions to maintain an old copy of the state and to undo the incorrect state change. In Figure 11-5, the store to *b−>sum* is undone by storing the old value back into *b−>sum.* The old value has been maintained here by allocating a new register for the new value (*tmp4*) and preserving the old value in *tmp3.* In some cases, it would be necessary to introduce new instructions to preserve the old value. For example, if *b−>ptr* had not been loaded, it would be necessary to introduce this load to obtain the old value.

By adding compensation code, software is performing the function of the hardware history buffer described in Section 5.1.3, except that software is more general. Software can restore nearly any processor and system state, not just registers. Furthermore, software can recognize that some items of state do not have to be restored on a branch misprediction. In Figure 11-5, the code to access *b−>sum* and to compute the value *b−>sum + a[i]* do not have to be undone because they affect temporary, intermediate values. Software can recognize this and not introduce the extra cycles to undo these intermediate operations, whereas a history buffer would take the time to restore all old values.

Compensation code adds instructions that can reduce the effectiveness of trace scheduling. In fact, if the compensation code adds more cycles than scheduling removes, the performance of trace-scheduled code is worse than the performance of the original code. Besides using a minimum number of instructions to perform compensation, there are two other ways that the trace-scheduling software can avoid inefficiency. The first is to use accurate branch predictions, so that the compensation code is added in sections of code that are not executed very much. The second is to use machine parallelism in performing compensation. If compensation can use otherwise free processor resources and can be performed early enough to avoid dependencies, it can be performed without adding cycles.

The requirement for efficient compensation code has an important bearing on the types of applications and processor architectures that are amenable to trace scheduling:

- Branch predictions should be accurate. If branch predictions are not accurate, the traces that are most efficiently scheduled may not be executed very often, and the sections of the code with the most compensation code may be executed frequently. In this respect, scientific applications are more amenable to trace scheduling than general-purpose applications.

- Branches should be infrequent with respect to other instructions. The higher the frequency of branches, the less likely it is that the code contains a few important traces that can be scheduled effectively. In this respect, again, scientific applications are more amenable to trace scheduling than general-purpose applications.

- The compensation code should not be difficult to construct. This argues for a regular instruction set, with nondestructive operations to preserve old state (such as the three-address instructions of most RISC architectures). Also, this argues against compound instructions (such as those of CISC architectures) that modify many items of state and argues against instructions with complicated side effects. In a processor without the appropriate properties, the scheduler must introduce many instructions to copy state on frequently executed traces as well as many instructions to restore state off of the traces.

- The more parallelism that hardware supports, the more likely it is that compensation can be performed with no penalty. Trace scheduling can help uncover instruction parallelism and make effective use of parallel hardware, but also relies on parallel hardware to reduce the cost of compensation. In machines tuned to applications that have a small amount of instruction parallelism, a large amount of machine parallelism is unnecessary, and compensation is less likely to be free.

It is no accident that trace scheduling has been used with VLIW processors for scientific applications. VLIW processors have a large amount of machine parallelism, a simple instruction set (in concept) and a simple hardware structure. Scientific applications justify the large amount of parallel hardware and have relatively predictable branch behavior with a large amount of computation per branch. The properties of these processors and applications make trace scheduling an appropriate technique for scientific applications. In general-purpose applications, trace scheduling has more limited usefulness.

11.1.3 Trace Scheduling an Entire Program

So far we have examined the scheduling of a single trace, but a program consists of many traces to be scheduled. To schedule a program, the scheduler considers one procedure at a time. Within a given procedure, the scheduler selects traces for scheduling according to the likelihood of execution, using predicted outcomes of the conditional branches in the procedure. To start, the scheduler performs trace scheduling on the trace that is most likely to execute. Compensation code is added off of the trace during this process. Following this, the next likely trace is selected for scheduling. This code includes any compensation code that may have been added during prior scheduling, so the compensation code also is scheduled. The process is repeated for each trace in turn. This approach allocates resources first to the *major traces*, or those traces most likely to be executed, and places most of the compensation code off of the major traces.

Each procedure has 2^n traces, where n is the number of conditional branches in the procedure. As the number of branches grows, the number of traces gets quite

---

large–a problem that is especially severe in general-purpose applications. Furthermore, the scheduling time and the amount of compensation code tend to grow as the number of branches grow. These effects combine to make the scheduling time and the size of the scheduled program grow quite rapidly with the number of branches. To limit scheduling time and to reduce the program growth, trace scheduling can be abandoned in favor of basic-block scheduling for the less likely traces. However, there are no obvious criteria for selecting the point at which to abandon trace scheduling.

11.1.4 Correctness of Trace Scheduling

The feasibility of trace scheduling depends on a relaxed definition of program correctness. The scheduled code is allowed to make temporary modifications to the program and system state that are different than those made by the original program, as long as the final modifications are exactly those that would have been made by the original program. This definition of correctness is essential because trace scheduling anticipates the execution of instructions. When instructions are incorrectly anticipated, the program must be able to discard the incorrect results and recover.

However, trace scheduling can encounter subtle problems arising from exceptions in moved instructions. An example of this appears in the trace-scheduled code of Figure 11-5, where the load of the *b–>sum* value has been moved to the preceding basic block. Consider that this instruction is moved in the same way in the following code fragment:

```
a[i] = a[i] + 1;
if ( b <> nil ) {
    b->sum = b->sum + a[i];
}
```

Here, the load of *b–>sum* is guarded by the *if* statement to insure that the *b* pointer is valid before the load executes. If the load is moved before the *if* branch, it is likely that the program will emit an invalid load. This will, at best, create an unexpected program error and, at worst, modify the system state unintentionally because the load has a side effect.

It is impossible–in general-purpose applications–for the scheduler to anticipate all possible effects of all moved instructions and to generate the appropriate compensation code. In some cases, compensation itself might be impossible–the processor cannot retract a character sent to a peripheral device, for example. Also, it is impossible for the scheduler to detect branches that guard against exceptional conditions so that it can prevent the movement of troublesome instructions. Even if the scheduler can do this in simple cases (in the most recent example, the programmer's intent is relatively obvious), the scheduler is still exposed to cases where the branches simply test other conditions that imply a valid operation. Branches may appear completely unrelated to the operations they guard. In practice, the scheduler must be conservative in moving code across branch boundaries, moving code only when it is absolutely safe to do so. As a result, the benefit of trace scheduling is reduced.

We have seen several problems with trace scheduling that argue against its use: the complexity and cost of compensation code; the potential for large code size and scheduling time; and the possibility of program errors caused by code motion that reduces the scope and benefit of possible code movement. Moreover, each of these problems is directly related to the number of branches in the code. The scheduler can attempt to avoid these difficulties by moving code only when it is "safe." Using this strategy, the scheduler moves instructions between basic blocks only when they do not require compensation and when they cannot create exceptions. However, compilers for scalar RISC processors use the same strategy to find instructions to fill the delay slots of branches. Measurements on the Stanford MIPS processor show that one safe instruction can be found to fill a branch-delay slot about 70% of the time, and two instructions can be found about 25% of the time [McFarling and Hennessy 1986]. If these measurements are representative–as they likely are–there are not many instructions that can be moved safely. This strategy is effective for the modest goal of filling a branch-delay slot, but not for the aggressive goal of uncovering significant instruction parallelism. The strength of trace scheduling lies in its ability to move code across branch boundaries, but this reduces trace scheduling to a special-purpose technique. In general-purpose applications, the large number of branches imposes many barriers to effective trace scheduling.

11.2 LOOP UNROLLING

Figure 11-6 shows the execution of a simple *while* loop that has been scheduled using list scheduling. The code shown for the loop has been optimized so that the index variable *i* has been removed. The test for *i* being equal to 64, to end the loop, has been replaced by a test for *a_ptr* being equal to *end_ptr*. This reduces the number of instructions in the loop, because it eliminates the instructions to update *i* and to perform the address computation for *a[i]*. On a processor with in-order issue, each iteration takes four cycles to execute.

The schedule for each loop iteration in Figure 11-6 does not use many processor resources, and each iteration has independent operations. Consequently, there is an opportunity to speed up the execution of this loop, but the loop as shown is not amenable to trace scheduling. Although the loop is a single basic block, it cannot be merged with adjacent basic blocks because instructions inside of the loop are to be executed once per iteration. If an instruction from outside of the loop is moved into the loop, it is executed too many times. If an instruction from inside of the loop is moved out of the loop, it is executed too few times. Instructions within the loop could be moved between iterations, but the trace scheduler does not have access to instructions of different iterations, since the loop consists of a single basic block.

```
                    for ( i=0 ; i<64 ; i++ ) {
                        sum += a[i];
                    }

        LOOP:  1:  load      tmp1,(a_ptr)          ;*a_ptr
               2:  add       a_ptr,a_ptr,4         ;a_ptr++
               3:  clt       cont,a_ptr,end_ptr    ;a_ptr<end_ptr?
               4:  add       sum,sum,tmp1          ;sum+=*a_ptr
               5:  jmpt      cont,LOOP             ;loop unless done
```

Decode				Execute					Writeback		Cycle
I0	I1	I2	I3	ALU1	ALU2	SHF	LS	BRN	R1	R2	
1:load	2:add	3:clt	4:add								1
		3:clt	4:add	2:add			1:load				2
			4:add		3:clt				2		3
5:jmpt				4:add				5:jmpt	3	1	4
1:load	2:add	3:clt	4:add						4		5
		3:clt	4:add		2:add		1:load				6
			4:add	3:clt					2		7
5:jmpt					4:add				3	1	8
								5:jmpt	4		9

Figure 11-6. Two Iterations of a List-Scheduled Loop

11.2.1 Unrolling to Improve the Loop Schedule

To perform trace scheduling on the loop shown in Figure 11-6, the loop must be *unrolled*. When a loop is unrolled, the instructions for two or more loop iterations are written explicitly. In Figure 11-7, the instructions for two iterations of the loop in Figure 11-6 are written in a single loop. The result is a larger loop that executes half as many times as the loop in Figure 11-6. Except for the loop termination test (the *clt* and *jmpt* instructions), the same number of operations is performed by both loops.

Unrolling transforms a loop into a trace and exposes instructions from different iterations to the software scheduler [Weiss and Smith 1987]. Figure 11-8 shows how list scheduling the unrolled loop improves performance. The loop now executes two iterations in four cycles, double the original loop performance of one iteration every four cycles. Because the loop executes many times, this can yield a large reduction in the program's overall execution time.

11.2.2 Unrolling with Data-Dependent Branches

With the loop in the previous section, the number of loop iterations is known to the compiler. This knowledge is important, because it helps the compiler relate itera-

---

```
                      for ( i=0 ; i<64 ; i++ ) {
                          sum += a[i];
                      }
```

```
LOOP:1:   load      tmp1,(a_ptr)          ;*a_ptr
     2:   add       a_ptr,a_ptr,4         ;a_ptr++
     3:   add       sum,sum,tmp1          ;sum+=*a_ptr
     4:   load      tmp1,(a_ptr)          ;*a_ptr
     5:   add       a_ptr,a_ptr,4         ;a_ptr++
     6:   add       sum,sum,tmp1          ;sum+=*a_ptr
     7:   clt       cont,a_ptr,end_ptr    ;a_ptr<end_ptr?
     8:   jmpt      cont,LOOP             ;loop unless done
```

Figure 11-7. Loop of Figure 11-6 Unrolled Twice

```
LOOP:1:   load      tmp1,(a_ptr)          ;*a_ptr
     2:   add       a_ptr,a_ptr,4         ;a_ptr++
     4:   load      tmp2,(a_ptr)          ;*a_ptr
     5:   add       a_ptr,a_ptr,4         ;a_ptr++
     7:   clt       cont,a_ptr,end_ptr    ;a_ptr<end_ptr?
     3:   add       sum,sum,tmp1          ;sum+=*a_ptr
     6:   add       sum,sum,tmp2          ;sum+=*a_ptr
     8:   jmpt      cont,LOOP             ;loop unless done
```

Decode				Execute					Writeback		
I0	I1	I2	I3	ALU1	ALU2	SHF	LS	BRN	R1	R2	Cycle
1:load	2:add	4:load	5:add								1
		4:load	5:add	2:add			1:load				2
7:clt	3:add	6:add	8:jmpt		5:add		4:load		2		3
		6:add	8:jmpt	7:clt	3:add				5	1	4
				6:add				8:jmpt	7	4	5
									3	6	6

Figure 11-8. List Scheduling Unrolled Loop of Figure 11-7

tions in the unrolled loop to iterations in the original loop. Since there are 64 iterations in the original loop, there are 32 iterations in the loop when it is unrolled twice.

However, the number of loop iterations is not always known to the compiler. For example, Figure 11-9 shows another way to write the loop of Figure 11-6. This style is sometimes used in the C programming language for efficiency, because it eliminates the variable i and the operations on this variable, just as the compiler optimizations did for the previous version of the loop. Unfortunately, however, the loop test is converted to a data-dependent branch: that is, a branch that depends on

---

```
            a_ptr = &(a[0]);
            end_ptr = a_ptr + 64 * sizeof(a[0]);
            while ( a_ptr<end_ptr ) {
                sum += *a_ptr;
                a_ptr += 1;
            }


    LOOP: 1:  load      tmp1,(a_ptr)          ;*a_ptr
          2:  add       a_ptr,a_ptr,4        ;a_ptr++
          3:  add       sum,sum,tmp1         ;sum+=*a_ptr
          4:  clt       cont,a_ptr,end_ptr   ;a_ptr<end_ptr?
          5:  jmpf      cont,EXIT            ;loop unless done
          6:  load      tmp1,(a_ptr)          ;*a_ptr
          7:  add       a_ptr,a_ptr,4        ;a_ptr++
          8:  add       sum,sum,tmp1         ;sum+=*a_ptr
          9:  clt       cont,a_ptr,end_ptr   ;a_ptr<end_ptr?
         10:  jmpt      cont,LOOP            ;loop unless done
    EXIT:
```

Figure 11-9. Loop with Data-Dependent Branch Unrolled Twice

data computed within the loop. This conceals the number of loop iterations from the compiler, and the compiler does not know that the total number of loop iterations is a multiple of two.

In code with data-dependent branches, the compiler can still unroll the loop, but now must include the loop test in each unrolled iteration. The unrolled loop is not a large basic block, but rather a trace of two or more adjacent basic blocks. In this case, the scheduler must rely on trace scheduling to schedule across iterations. If the scheduler now moves code across iterations, it must add compensation code off of the associated loop exit points. This complicates scheduling and reduces the benefit of loop unrolling.

11.3 SOFTWARE PIPELINING

Figure 11-8 illustrates how unrolling a loop twice exposes instructions from different loop iterations to the scheduler and helps improve performance. Theoretically, unrolling the loop more than twice should expose even more parallelism and improve performance even more. However, the hardware we have been using does not support this amount of parallelism. Because hardware resources are limited, it is not possible to have more than two loop iterations in execution at any given time, and thus there is no advantage to unrolling the loop more than twice. This observation holds in general when loop operations have short latencies. In this case, each loop iteration executes at a rate of one or more instructions per cycle–using one or more functional units and result buses per cycle–and it is often impossible to have more than two loop iterations in execution without exhausting processor resources.

In loops that execute less than one instruction per clock cycle, however, the hardware can support the execution of more than two adjacent loop iterations. Figure 11-10 shows an example of such a loop. This loop executes at a rate of less than one instruction per cycle because it contains a floating-point multiply. If we assume that this is a single-precision multiply, the latency of the operation is four cycles (in our hardware model). Figure 11-11 shows this loop scheduled after being unrolled both two times and four times; the numbers to the right of the code identify the iteration to which each instruction belongs, for clarity. Note that the scheduler has created two versions of *a_ptr* for the loop unrolled twice (*a_ptr1* and *a_ptr2*) and four versions of *a_ptr* for the loop unrolled four times (*a_ptr1* through *a_ptr4*). This is an example of software renaming. Using different instances of *a_ptr* preserves the value of *a_ptr* for a given iteration, to complete the store at the end of the iteration, while allowing *a_ptr* to be updated for the next iteration.

Figure 11-12 shows the execution sequences of the two loops shown in Figure 11-11, in a processor with in-order issue (the floating-point multiplier occupies the functional-unit column used for the shifter in previous examples). This schedule assumes a pipelined multiplier, allowing a multiply to be initiated on each cycle. The loop unrolled twice executes in 9 cycles, so it executes the original loop at the rate of 4.5 cycles per iteration. The loop unrolled four times executes in 10 cycles, so it executes the original loop at the rate of 2.5 cycles per iteration. There is an obvious advantage to unrolling the loop more than twice. In fact, the loop can be unrolled six times to achieve an execution rate of 2 cycles per iteration before the limitation of a single load/store unit prevents any further advantage.

11.3.1 Pipelining Operations from Different Loop Iterations

Figure 11-12 illustrates that scheduling an unrolled loop effectively pipelines operations from different loop iterations during the execution of the unrolled loop. For

```
float a[64], b;
for ( i=0 ; i<64 ; i++ ) {
    a[i] = b * a[i];
}
```

Unscheduled Code for Loop

```
LOOP: load    tmp1,(a_ptr)         ;*a_ptr
      fmul    tmp1,b,tmp1          ;b*a[i]
      store   tmp1,(a_ptr)         ;assign a[i]
      add     a_ptr,a_ptr,4        ;a_ptr++
      clt     cont,a_ptr,end_ptr   ;a_ptr<end_ptr?
      jmpt    cont,LOOP            ;loop unless done
```

Figure 11-10. A Loop with Long Operation Latencies

```
                       float a, b;
                       for ( i=0 ; i<64 ; i++ ) {
                           a[i] = b * a[i];
                       }
```

Unrolled two times

```
LOOP: 1:  add     a_ptr2,a_ptr1,4    ;a_ptr++           2
      2:  load    tmp1,(a_ptr1)      ;*a_ptr            1
      3:  load    tmp2,(a_ptr2)      ;*a_ptr            2
      4:  fmul    tmp1,b,tmp1        ;b*a[i]            1
      5:  fmul    tmp2,b,tmp2        ;b*a[i]            2
      6:  store   tmp1,(a_ptr1)      ;assign a[i]       1
      7:  add     a_ptr1,a_ptr2,4    ;a_ptr++           1
      8:  clt     cont,a_ptr1,end_ptr;a_ptr<end_ptr?
      9:  store   tmp2,(a_ptr2)      ;assign a[i]       2
     10:  jmpt    cont,LOOP          ;loop unless done
```

Unrolled four times

```
LOOP: 1:  add     a_ptr2,a_ptr1,4    ;a_ptr++           2
      2:  load    tmp1,(a_ptr1)      ;*a_ptr            1
      3:  add     a_ptr3,a_ptr2,4    ;a_ptr++           3
      4:  load    tmp2,(a_ptr2)      ;*a_ptr            2
      5:  add     a_ptr4,a_ptr3,4    ;a_ptr++           4
      6:  load    tmp3,(a_ptr3)      ;*a_ptr            3
      7:  fmul    tmp1,b,tmp1        ;b*a[i]            1
      8:  fmul    tmp2,b,tmp2        ;b*a[i]            2
      9:  load    tmp4,(a_ptr4)      ;*a_ptr            4
     10:  fmul    tmp3,b,tmp3        ;b*a[i]            3
     11:  fmul    tmp4,b,tmp4        ;b*a[i]            4
     12:  store   tmp1,(a_ptr1)      ;assign a[i]       1
     13:  add     a_ptr1,a_ptr4,4    ;a_ptr++           1
     14:  store   tmp2,(a_ptr2)      ;assign a[i]       2
     15:  clt     cont,a_ptr1,end_ptr;a_ptr<end_ptr?
     16:  store   tmp3,(a_ptr3)      ;assign a[i]       3
     17:  store   tmp4,(a_ptr4)      ;assign a[i]       4
     18:  jmpt    cont,LOOP          ;loop unless done
```

Figure 11-11. Unrolling the Loop of Figure 11-10

example, when the loop is unrolled four times, an iteration from the original loop is initiated on every cycle for the first four cycles of the unrolled loop, as the pipeline is filled. An iteration ends on every cycle for the last four cycles of the loop, as the pipeline is drained. The loop operations do not require this draining and filling of the pipeline–draining and filling are simply an artifact of loop unrolling.

There is another way to pipeline operations from different loop iterations, without unrolling the loop. Figure 11-13 shows the execution of a loop that has been assembled with instructions from different iterations of the original loop, using a technique called *software pipelining* [Charlesworth 1981, Rau and Glaeser 1981, Lam

Figure 11-12. Execution Schedules for the Loops of Figure 11-11

Unrolled two times

Cycle	ALU1	ALU2	FMUL	LS	BRN
1	add/2			load/1	
2				load/2	
3			fmul/1		
4			fmul/2		
5					
6					
7	add/1			store/1	
8		cmpr		store/2	
9					brn

Unrolled four times

Cycle	ALU1	ALU2	FMUL	LS	BRN
1	add/2			load/1	
2	add/3			load/2	
3	add/4		fmul/1	load/3	
4			fmul/2	load/4	
5			fmul/3		
6			fmul/4		
7	add/1			store/1	
8		cmpr		store/2	
9				store/3	
10				store/4	brn

Figure 11-13. Execution of a Software-Pipelined Loop

Decode:

I0	I1	I2	I3	Cycle
load/5	add/5	fmul/4	clt/4	n
store/2	jmpt/4			
load/6	add/6	fmul/5	clt/5	
store/3	jmpt/5			
load/7	add/7	fmul/6	clt/6	
store/4	jmpt/6			
load/8	add/8	fmul/7	clt/7	
store/5	jmpt/7			
load/9	add/9	fmul/8	clt/8	
store/6	jmpt/8			
load/10	add/10	fmul/9	clt/9	
store/7	jmpt/9			

Execute / Writeback:

ALU1	ALU2	FMUL	LS	BRN	R1	R2	Cycle
add/5	clt/4	fmul/4	load/5		x		n+1
			store/2	jmpt/4	x	x	n+2
add/6	clt/5	fmul/5	load/6		x	x	n+3
			store/3	jmpt/5	x	x	n+4
add/7	clt/6	fmul/6	load/7		x	x	n+5
			store/4	jmpt/6	x	x	n+6
add/8	clt/7	fmul/7	load/8		x	x	n+7
			store/5	jmpt/7	x	x	n+8
add/9	clt/8	fmul/8	load/9		x	x	n+9
			store/6	jmpt/8	x	x	n+10
add/10	clt/9	fmul/9	load/10		x	x	n+11
			store/7	jmpt/9	x	x	n+12

1988, 1989]. The principal difference between software pipelining and loop unrolling is that software pipelining is not constrained to pipeline operations from adjacent loop iterations as is loop unrolling (we will see in a moment how this is accomplished). The number following an instruction in Figure 11-13 denotes the iteration of the original loop to which an instruction in the software-pipelined loop belongs. We will ignore for the moment how the *add* instruction can be advanced to the fifth iteration before the *store* enters the second iteration, and focus rather on the computation starting in the fifth iteration of the original loop.

During the first iteration of the software-pipelined loop shown in Figure 11-13, the processor issues the address addition for the fifth iteration of the original loop. The *load* using this address (from the sixth iteration of the original loop) is issued during the second iteration, and the *fmul* using the loaded data is issued during the third iteration. The *fmul* is in execution during the fourth iteration, and, finally, the data is stored during the fifth iteration. The dependencies in a given iteration of the original loop are satisfied by distributing instructions over many iterations of the software-pipelined loop, so that the instructions from different iterations can be overlapped.

Software pipelining can overlap instructions from widely separated iterations, rather than being constrained to schedule operations from several adjacent iterations as is the case with loop unrolling. This means that the loop pipeline can execute at the optimum rate, rather than being drained and filled at regular intervals. The software-pipelined loop in Figure 11-13 executes at a rate of two iterations per cycle, whereas the same loop must be unrolled six times to achieve this rate with basic-block scheduling. Because of the high degree of unrolling required, software pipelining achieves this throughput with fewer instructions.

Moreover, software pipelining can uncover more instruction parallelism. If the hardware in Figure 11-13 had two load/store pipelines and a wider decoder, software pipelining could use both load/store pipelines on every cycle and achieve a throughput of one loop iteration per cycle. With loop unrolling, the scheduler could not move a store earlier than the seventh cycle of an iteration (see Figure 11-12) and would run out of loads six cycles before the end of an iteration. Thus, with loop unrolling, the scheduler is not able to exploit machine parallelism to the extent that software pipelining is (unless the scheduler takes the extreme measure of unrolling the loop 64 times!).

11.3.2 Software-Pipelining Techniques

The goal of software pipelining is to execute loop iterations at the highest possible rate. This means that, after initiating a given loop iteration, the next loop iteration should be initiated at the earliest opportunity. The *initiation interval* is the time (in terms of processor cycles) between the initiation of a given iteration and the initiation of the next iteration. Scheduling for a minimum initiation interval is somewhat counter to the goals of scheduling as they have been presented so far. Basic-block scheduling would attempt to minimize the execution time of a single iteration. However, this does not insure that the next iteration is initiated as soon as possible, and might actually delay the initiation of the next iteration. For example, to optimize the execution time of a single iteration, the scheduler might assign resources that are needed to start the next iteration, thereby increasing the initiation interval and overall execution time (see Figure 11-14). The scheduler should instead assign resources in favor of initiating subsequent loop iterations. The scheduler also should schedule

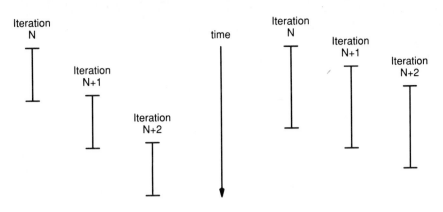

Minimum Iteration Time

Figure 11-14.   Scheduling for Minimum Iteration Time Versus Minimum Initiation Interval

operations within a loop in favor of computing the values needed by subsequent iterations, again so that these iterations can begin as soon as possible.

There are several different approaches to software pipelining. Charlesworth [1981] and Rau and Glaeser [1981] discuss software-pipelining algorithms that work in special cases or with special hardware. Aiken and Nicolau [1988] propose unrolling loops and searching for a pipelining pattern in the scheduled code. This section outlines the software-pipelining algorithms developed by Lam [1988, 1989]. These algorithms are useful for a variety of loops and are easily understood from the principles presented so far.

During software pipelining, the scheduler detects resource conflicts among instructions in different iterations (but in the same pipelined loop) using a *modulo reservation table*. Figure 11-15 shows the relationship between the modulo reservation table and the reservation table of a single loop iteration. The modulo reservation table has entries for all processor resources, as does the reservation table (only the functional units are shown in Figure 11-15). However, the modulo reservation table differs in that it has entries only for the cycles in the initiation interval, rather than the cycles in the entire schedule. The resource utilization of an entire iteration in the original loop is compressed into the modulo reservation table using the modulus function: if a resource is used by an instruction during cycle n in the reservation table, then the same resource is used in cycle $n \bmod i$ in the modulo reservation table, where i is the initiation interval.

This definition of the modulo reservation table is based on the observation that, to achieve the given initiation interval, the resource utilization of all overlapped iterations must not conflict within the interval, implying that instructions will not con-

---

Reservation Table for an Iteration

ALU1	ALU2	FMUL	LS	BRN
add			load	
	clt			
		fmul		jmpt
			store	

Modulo Reservation Table (initiation interval = 2)

ALU1	ALU2	FMUL	LS	BRN
add	clt		load	
		fmul	store	jmpt

Figure 11-15. The Modulo Reservation Table

flict within an iteration of the pipelined loop. If an instruction uses a resource in cycle n in the reservation table, and n is greater than the initiation interval i, then the same instruction *from another iteration* will use the resource during cycle $n \bmod i$ of an iteration in the pipelined loop. The resource utilization of a single iteration in the original loop must be fit into the modulo reservation table so that all instructions that appear in the original iteration can also appear in the initiation interval when the loop is pipelined, although, in the pipelined loop, these instructions are usually from different iterations of the original loop.

The scheduling for software pipelining is similar to basic-block scheduling, except that the modulo reservation table is used instead of the reservation table. The resulting schedule is not the same as a schedule for a single iteration, because the goal is to minimize the initiation interval rather than to minimize the execution time of a single iteration. Because of this, the resource reservation table for an iteration in Figure 11-15 is not the same table that list scheduling would construct for a single iteration of the same loop. (The reservation table and modulo reservation table in Figure 11-15 also do not precisely reflect resource utilization during execution as shown in Figure 11-13. The hardware issues instructions after they are free of dependencies, making the execution order slightly different than the software schedule.)

Software pipelining encounters a circularity that is illustrated by the role of the modulo reservation table. The size of the modulo reservation table is determined by the initiation interval, but the initiation interval depends on the success of scheduling without having conflicts in the modulo reservation table. Obviously, the shorter the initiation interval, the more likely it is that operations in the loop will conflict for resources, because the resources are available for fewer cycles. However, it is not necessarily true that increasing the initiation interval will remove resource conflicts. For example, two uses of a resource that are separated by two cycles do not conflict

Superscalar Microprocessor Design

when the initiation interval is two cycles (because they occur in different cycles modulo two), but do conflict when the initiation interval is three cycles (because they occur in the same cycle modulo three).

Though the initiation interval cannot be known when scheduling begins, the lower and upper bounds of the initiation interval can be known, because these bounds do not depend on the success of the scheduler. The lower bound is set either by the dependencies on values computed in one iteration and used in a subsequent iteration, or by the resource utilization of an iteration. If an iteration computes a value used in a subsequent iteration, the initiation interval cannot be so short that the value has not been computed by the time it is needed. Also, there must be enough cycles in the initiation interval to provide for the resource utilization of all instructions in an iteration. The lower bound is set by the largest of these two constraints. In the example we have been using, the *add* instruction computes an address used by the *load* instruction in the next iteration. However, this only constrains the initiation interval to one cycle. The utilization of the load/store unit imposes a more severe restriction, because a single iteration uses this resource twice and there is only one resource. The utilization of the decoder imposes the same lower bound, because there are six instructions in the loop and only four decoder positions. Thus, there must be at least two cycles in the initiation interval to avoid resource conflicts. Again, this initiation interval is a lower bound—it does not guarantee that there will be no resource conflicts.

The upper bound on the initiation interval is simply the schedule of a single loop iteration using basic-block scheduling. With this initiation interval, there is no software pipelining.

The circularity in software pipelining caused by the interdependence of the initiation interval and the scheduling constraints is broken by trial and error. The scheduler first attempts to schedule with the initiation interval set at the lower bound. If this is unsuccessful, the scheduler increases the initiation interval by one cycle and repeats the attempt. This process is repeated until the loop is successfully pipelined or the upper bound is reached. If the upper bound is reached, the schedule must be successful, but the loop is not pipelined. Typically, only a small number of attempts is required to successfully pipeline the loop.

Figure 11-16 shows the dataflow graph of a single iteration of the example loop. This graph has an example of a *loop-carried* dependency: the *add* instruction from a given iteration must complete before the *load* of the next iteration can begin. This dependency appears because software pipelining schedules across the loop branches and must be aware of dependencies that cross these branches. Also, in the original loop, there is an antidependency between the *add* instruction and the *store* instruction, because the address for the *store* cannot be changed until the *store* is issued. This dependency can be removed using software renaming (the technique will be described in Section 11.3.4). If this antidependency were not removed from the

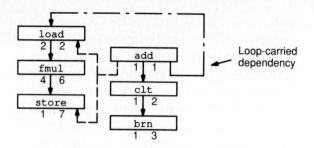

Figure 11-16. Dataflow Graph of a Loop Iteration

loop, the loop could not be pipelined. The *add* instruction would have to wait on the completion of the *store* at the bottom of the loop, and the *load* for the next iteration could not begin until the *add* were completed. In the current discussion, we will assume that the antidependency has been removed.

Software pipelining proceeds in two steps. The first step is to schedule the instructions involving loop-carried dependencies. Since this is performed first, it gives priority for processor resources to instructions that are required to initiate subsequent iterations of the loop. If there are no loop-carried dependencies, this step is not necessary. In our example, the *add* and the *load* instructions are scheduled first, and they are scheduled to execute in the same cycle, with the *add* following the *load* in the code. This satisfies the antidependency between the *add* and the *load*, because the *add* follows the *load*, and it completes the *add* as early as possible for the *load* in the next iteration. After these instructions are scheduled, they are merged into a single node in the dependency graph. The resource utilization of the merged node is the union of the utilization of the merged instructions (see Figure 11-17). The loop-carried dependencies are satisfied by the schedule of the merged node, and these dependencies need not be considered further.

The next step in software pipelining is to schedule the new loop containing the merged node. The goal of this step is to schedule instructions so that resource usage

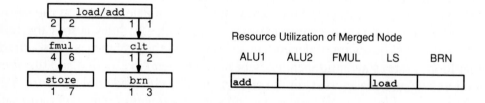

Figure 11-17. Merged Node After Interiteration Scheduling

Superscalar Microprocessor Design

fits within the initiation interval. Figure 11-18 shows the scheduling steps, primarily to illustrate the use of the modulo reservation table. In the sixth step, the merged node is scheduled. This node could be scheduled in the second cycle of the initiation interval while satisfying data dependencies, except that the *load* instruction in the merged node would conflict with the *store* instruction. Because of the conflict, the merged node is moved up in the schedule to the first cycle of the initiation interval. This resource conflict has no effect on the final instruction ordering–the *load* and *add* instructions are scheduled first in any case–but this example illustrates the scheduling of merged nodes and the detection of resource conflicts that can interfere with the pipelining of the loop.

In this example, the scheduler has pipelined the loop with the minimum initiation interval allowed by the hardware. As discussed before, this scheduling could have failed, in which case the scheduler would have increased the initiation interval and attempted the schedule again until a successful schedule was found. Figure 11-15 shows the schedule of one loop iteration.

11.3.3 Filling and Flushing the Pipeline: The Prologue and Epilogue

The body of a software-pipelined loop contains instructions from different iterations of the original loop. Within the pipelined loop, instructions near the top of the data-flow graph can be in iterations far beyond the iterations of instructions near the bottom of the graph. Before the loop begins executing, the loop pipeline must be filled by issuing the earlier iterations of the instructions near the top of the graph. The scheduler accomplishes this by placing the appropriate instructions in a loop *prologue*. The prologue is executed once before the loop begins. Also, when the loop test detects the final iteration, some instructions may be in earlier iterations. When this happens, the loop pipeline must be drained by issuing the instructions to finish any incomplete iterations. The scheduler accomplishes this by placing the ap-

1:

ALU1	ALU2	FMUL	LS	BRN
			store	

2:

ALU1	ALU2	FMUL	LS	BRN
			store	brn

3:

ALU1	ALU2	FMUL	LS	BRN
	clt			
			store	brn

4:

ALU1	ALU2	FMUL	LS	BRN
	clt			
		fmul	store	brn

5: (merged node not in ready set)

ALU1	ALU2	FMUL	LS	BRN
	clt			
		fmul	store	brn

6:

ALU1	ALU2	FMUL	LS	BRN
add	clt		load	
		fmul	store	brn

Figure 11-18. Entries in Modulo Reservation Table During Software Pipelining

propriate instructions in a loop *epilogue*. The epilogue executes once after the loop terminates.

Figure 11-19 shows the execution of the loop prologue and epilogue for our example. In this example, the *load* and *add* instructions are three iterations ahead of the *store* instruction within the pipelined loop, and the *clt*, *fmul*, and *jmpt* instructions are two iterations ahead. The prologue should therefore start three iterations of the *load* and *add* instructions and two iterations of the other instructions. Software knows in this case that the loop will execute more than three times and can omit the loop test and branch within the prologue as shown in Figure 11-19. The other instructions are started as required. Also, within the pipelined loop, the *store* instruction is two iterations behind the loop branch. The epilogue must therefore complete the two iterations of the *store* that remain when the loop terminates.

When the pipelined loop executes its final iteration in Figure 11-19, the loop issues the *load* and *add* instructions for one iteration beyond the end of the loop. These are spurious instructions caused by scheduling instructions in advance of the

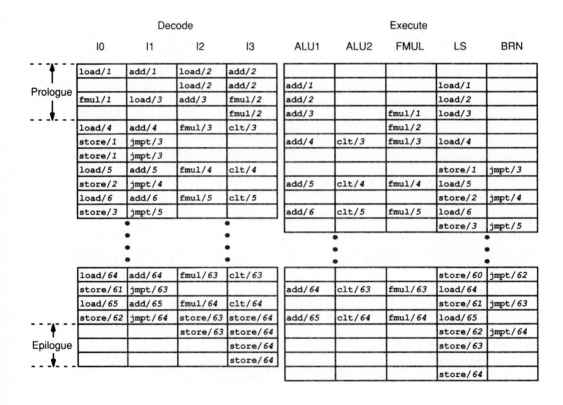

Decode				Execute				
I0	I1	I2	I3	ALU1	ALU2	FMUL	LS	BRN
load/1	add/1	load/2	add/2					
		load/2	add/2	add/1			load/1	
fmul/1	load/3	add/3	fmul/2	add/2			load/2	
			fmul/2	add/3		fmul/1	load/3	
load/4	add/4	fmul/3	clt/3			fmul/2		
store/1	jmpt/3			add/4	clt/3	fmul/3	load/4	
store/1	jmpt/3							
load/5	add/5	fmul/4	clt/4				store/1	jmpt/3
store/2	jmpt/4			add/5	clt/4	fmul/4	load/5	
load/6	add/6	fmul/5	clt/5				store/2	jmpt/4
store/3	jmpt/5			add/6	clt/5	fmul/5	load/6	
							store/3	jmpt/5
⋮		⋮		⋮			⋮	
load/64	add/64	fmul/63	clt/63				store/60	jmpt/62
store/61	jmpt/63			add/64	clt/63	fmul/63	load/64	
load/65	add/65	fmul/64	clt/64				store/61	jmpt/63
store/62	jmpt/64	store/63	store/64	add/65	clt/64	fmul/64	load/65	
		store/63	store/64				store/62	jmpt/64
			store/64				store/63	
			store/64				store/64	

Figure 11-19. Execution of the Loop Prologue and Epilogue

loop branch, and they can cause many of the same problems as spurious instructions generated by trace scheduling (see Section 11.1.4). In this example, software can avoid the spurious instructions by adjusting the loop test so that the loop terminates one iteration early (the epilogue then includes an *fmul* and another *store*). Another alternative that has been proposed for cases such as this is to allocate an extra location for the array so that the spurious *load* does not cause an invalid access, but this is a strange way to handle the problem.

The spurious instructions in our example are relatively easy to deal with, because software knows the number of loop iterations in advance. Furthermore, the loop test involves a short dependency path from the beginning of the loop, so there are only two instructions in iterations beyond the loop test. However, these characteristics do not apply to loops with data-dependent branches. For these loops, the scheduler cannot adjust the loop test to avoid spurious instructions because it does not know the number of loop iterations. Also, a loop test that is dependent on computation within the loop often is near the bottom of the loop dependency-graph. This means that the scheduler might schedule instructions several iterations beyond the loop test, in which case these instructions would have been issued when the last iteration is detected. If this were true, there would be many spurious instructions. For example, a software-pipelined loop that traverses a *nil*-terminated list might generate a reference with a *nil* pointer, because it may attempt to access the (nonexistent) item beyond the end of the list before it processes the last item in the list. However, as in trace scheduling, the scheduler does not actually allow unsafe execution. The *possibility* of unsafe execution, however, causes the scheduler to be conservative, reducing performance. For cases such as traversing a *nil*-terminated list, the possibility of unsafe execution restricts the overlap of loop iterations, so that the last iteration can be detected before the execution of any unsafe, spurious instructions beyond the last iteration. Reducing the amount of loop overlap reduces loop throughput. The longer it takes to determine the outcome of the loop branch, the longer the subsequent iterations must wait, and the greater the reduction in loop throughput.

11.3.4 Register Renaming in the Software-Pipelined Loop

We saw, at the beginning of this section, an example of register renaming to aid code motion within an unrolled loop. By changing the allocation of registers, the scheduler avoids storage conflicts between instructions in different iterations and increases the instruction parallelism within the unrolled loop. Renaming helps software pipelining the same way, removing storage conflicts between instructions in different iterations of a pipelined loop.

The software-pipelined loop shown in Figure 11-13 (page 219) has several examples of potential storage conflicts. These storage conflicts are not resolved by the schedule of the pipelined loop, and the conflicts must be removed by applying software renaming for the loop to achieve the execution rate shown.

Some of the storage conflicts in Figure 11-13 are apparent in the original loop. For example, there are antidependencies between the *add* that updates the address for the loop and the *load* and *store* instructions in the same loop. As discussed in Section 11.3.2, the need to preserve the address for the *store* can inhibit the address computation for the *load* in subsequent iterations. Having to wait for the *store* to complete before starting another iteration effectively eliminates the possibility of pipelining.

Other storage conflicts in Figure 11-13 are introduced by overlapping the loop iterations. These conflicts appear between instructions in different iterations of the original loop. For example, there is an output dependency between a *load* in an iteration and the *load* in the previous iteration. This dependency is a concern because the second *load* is issued before the *fmul* that uses the data from the first *load*. The *fmul* should not incorrectly receive the value from the second iteration. Interestingly, this problem is introduced by register-dependency hardware. A processor without dependency hardware (such as the VLIW processors for which software pipelining was developed) would deliver the current value of the register regardless of any new, pending updates. This would correctly deliver the result of the first *load* to the *fmul*, even with the intervening second *load*, because the second *load* would not have completed when the *fmul* was issued. A processor with dependency hardware detects that there is a new update to the register and stalls the *fmul* so that the incorrect value is delivered.

Dependency hardware creates a need for software to remove the output dependency from the pipelined loop. This does not argue against dependency hardware, however. A program that relies on hardware latencies to remove storage conflicts may not work if the latencies change. Dependency hardware makes software independent of hardware latencies. Furthermore, software renaming is important whether or not dependency hardware exists. Dependency hardware does not impose the requirement to perform software renaming: it simply provides more conditions for software to consider.

In an unrolled loop, software renaming is straightforward, because there are separate instructions for each unrolled iteration. The registers allocated for an iteration are changed just by changing the instructions in the iteration. In a software-pipelined loop, on the other hand, there are not necessarily enough instructions to perform the appropriate register allocation. Software must therefore explicitly unroll the pipelined loop to allow register allocation.

In the example of Figure 11-13, four loop iterations are initiated from the time that the address is computed until that address is used for the *store*. None of these four iterations can disturb the address value. Thus, the loop must be unrolled five times, so that four iterations can be initiated between the address *add* and the *store*. The allocation of registers across the iterations is as follows (iteration numbers designate unrolled iterations):

Iteration 1	Iteration 2	Iteration 3	Iteration 4	Iteration 5
`load  t1,a1`	`load  t2,a2`	`load  t3,a3`	`load  t4,a4`	`load  t5,a5`
`add   a2,a1`	`add   a3,a2`	`add   a4,a3`	`add   a5,a4`	`add   a1,a5`
`fmul  t5`	`fmul  t1`	`fmul  t2`	`fmul  t3`	`fmul  t4`
`clt   a5`	`clt   a1`	`clt   a2`	`clt   a3`	`clt   a4`
`store t3,a3`	`store t4,a4`	`store t5,a5`	`store t1,a1`	`store t2,a2`
	`jmpf  a1`			`jmp`

Since software knows the number of loop iterations in this example, it knows that the loop can be exited only at the second unrolled iteration (accounting for the iterations started in the prologue). Note that the exit test is complemented, because it must skip the other iterations to exit the loop. The jump at the bottom of the unrolled loop just returns to the top of the loop. If this loop had data-dependent branches, it would also have all of the attendant problems–plus one. There would have to be a different version of the epilogue for each iteration in the unrolled loop, because a different set of registers would be used to contain loop variables at each exit point.

Comparing the final results of software pipelining to the example of loop unrolling at the beginning of this section, we see that the software-pipelined loop must be unrolled five times to match the throughput of the nonpipelined loop which is unrolled six times. The degree of unrolling is determined by very different factors, however. If the store were not present in this loop–for example, if the array elements were simply multiplied together–only two unrolled iterations would be necessary to avoid output dependencies between instructions in different loop iterations.

Still, software pipelining can require a very high degree of unrolling. For example, one loop variable may have to be preserved across three iterations while another must be preserved across four iterations. In this case, the loop would have to be unrolled 12 times to allocate registers properly. Rather than incur this code expansion, it might be better in such cases to tolerate lower performance.

We began this section by introducing software pipelining as a better alternative to simple loop unrolling for loops that have long operation latencies, particularly loops with floating-point operations. Unfortunately, having long operation latencies also increases the probability that the software-pipelined loop will have to be unrolled to implement software renaming. This is not an indictment of software pipelining with respect to simple unrolling, because software pipelining can still achieve a higher execution rate for a given code expansion. The need to unroll the pipelined loop simply reduces the benefit of software pipelining in some cases.

11.4 GLOBAL CODE MOTION

Some of the code motion that is performed during trace scheduling (see Figure 11-4, page 208) involves very simple compensation code. If an instruction is moved up beyond a point where multiple paths converge into one, the instruction must be duplicated on all paths. Similarly, if an instruction is moved down beyond a point where a single path diverges into multiple paths, the instruction must be duplicated on all paths. These are examples of *global code motion*. Code is moved across basic-block boundaries, but only in ways that do not depend on knowing the outcomes of branches. Trace scheduling performs global code motion, but also performs other code motion requiring more complex compensation code. This section briefly considers other scheduling techniques that perform global code motion.

Foster and Riseman [1972] propose a scheduling technique called *percolation*. Percolation relies on converting common sequences of basic blocks into larger basic blocks (which Foster and Riseman call *chunks*). Figure 11-20 illustrates two possible block conversions. The first conversion shown in Figure 11-20 merges the code following an *if-then-else* statement into the two paths of the *if-then-else* (a similar conversion can be applied to a simple *if* statement). This creates larger basic blocks for scheduling along both paths of the *if-then-else*. The second conversion transforms a loop with a *break* so that code can move freely in the loop body.

The disadvantage of Foster and Riseman's percolation is that it performs global code motion by duplicating entire basic blocks instead of individual instructions. Nicolau [1985] proposes an approach that avoids this problem, which he also calls percolation scheduling. Nicolau's percolation scheduling always attempts to move instructions toward the beginning of the program. If an instruction is moved up beyond a point where paths converge, the instruction is duplicated along all paths. As the duplicated copies are moved along their respective paths by scheduling, they may later be moved to a common block. The percolation algorithm recognizes this and deletes all but one copy. For example, this can move an instruction into both paths of an *if-then-else* statement and then later move it above the statement.

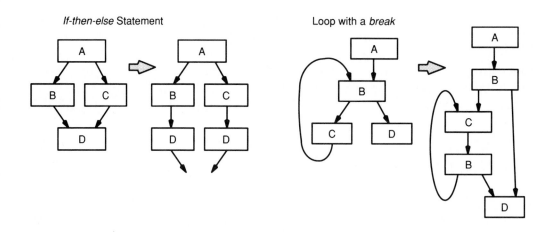

Figure 11-20. Forming Large Basic Blocks from Common Code Structures

Nicolau's percolation can also move branches, as shown in Figure 11-21. The percolation-scheduling transformations are intended as primitive operations used by a high-level scheduler. The high-level scheduler directs code motion and deals with program structures that are not comprehended by percolation, such as loops.

Lam [1988, 1989] has developed a unified approach to global code motion she calls *hierarchical reduction*. Hierarchical reduction systematically converts branching structures to nonbranching structures by applying the conversions shown in Figure 11-22. Both conversions in Figure 11-22 reduce the control flow graph to a straight-line graph (Figure 11-22 shows the conversion of a software-pipelined loop–a nonpipelined loop has the same conversion without the prologue and epilogue). After a conversion, the scheduler can apply basic-block scheduling to move

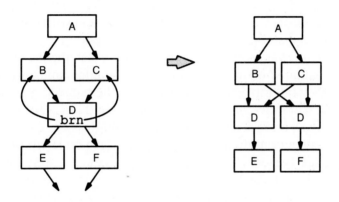

Figure 11-21. Percolating a Branch Instruction

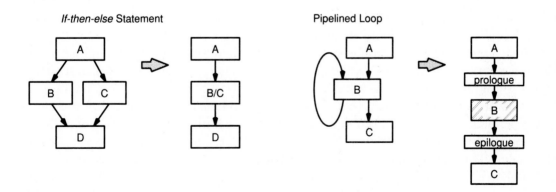

Figure 11-22. Merging Nodes for Hierarchical Reduction

instructions into both paths of an *if-then-else* statement or around a program loop. The branching structures are not changed in the actual code, but only in the flow graph used by the scheduler.

Hierarchical reduction converts both paths of an *if-then-else* into a single, combined node (we saw an example of a combined node created during software pipelining in Figure 11-17, page 224). The resource utilization of this combined node is the union of the resource utilization on both paths. The combined node also satisfies the worst-case precedence constraints of instructions on either path. If instructions from surrounding blocks are scheduled into the combined node, these instructions appear in both paths of the *if* statement in the final code. An instruction after the *if* statement can be moved before the statement, and vice versa, without duplicating the instruction.

In a simple *if* statement with no *else* clause, the null *else* path can still be merged (conceptually) with instructions on the other path, and hierarchical reduction still applies. However, an instruction moved into an *if* statement with a null *else* clause creates the need for an *else* branch where there was none in the original code. The moved instruction must appear on both paths of the *if* statement even though the *else* path originally contained no instructions. The scheduler can be made aware of the null *else* clause to avoid moving instructions into the reduced *if* node. In this case, the scheduler can still move instructions freely around the reduced node.

Hierarchical reduction converts a loop into a single node. If the loop is software pipelined before reduction, the prologue and epilogue code are placed in separate nodes preceding and succeeding the reduced loop node (software pipelining is most advantageous for innermost loops with long-latency operations, and other loops probably are not pipelined). An instruction is not allowed to move into or out of the loop body, because it would not execute the correct number of times; this is indicated by the cross-hatching on the reduced loop node in Figure 11-22. Instructions are prevented from moving into the loop using the same mechanism that prevents instructions from moving into an *if* statement with a null *else* clause. Instructions still can move freely around the loop, and instructions in the prologue and epilogue can be scheduled with instructions in surrounding nodes.

Hierarchical reduction is successively applied to the program, beginning with the innermost program loops and *if* statements and progressing outward, until the entire program has been scheduled. Before reduction, the innermost program loops are software pipelined, if necessary. If these loops contain *if* statements, the loops can be pipelined by first reducing the *if* statements. Each pipelined loop is then reduced, and the loop, along with its prologue and epilogue, is scheduled with surrounding code. This scheduling includes all instructions up to the boundaries of the next higher-level branching structure, such as the point where two paths of an *if* statement diverge or converge or the top or bottom of the next outer loop. Following this, the scheduled code is reduced to a single, combined node. Reduction and scheduling is then repeated for all other code contained within the next higher-level branching structure. When all such code has been scheduled and reduced in this manner, the higher-level branching structure itself is reduced. This process continues, with repeated application of scheduling and reduction, until the entire program has been reduced to a single node. Higher-level loops can be software pipelined, but this is rarely advantageous because the resource utilization of higher-level loops does not allow much overlap between loop iterations, and inner loops dominate the execution time anyway.

11.5 OUT-OF-ORDER ISSUE AND SCHEDULING ACROSS BRANCHES

We end this chapter by examining the most fundamental trade-off in the design of a superscalar microprocessor: whether or not to perform speculative, out-of-order issue. Within basic blocks, software scheduling is clearly superior to out-of-order issue, because software is able to examine many different instruction orderings and choose the best (or nearly best) one. Hardware cannot examine the entire basic block at once–at least not without enormous expense. Hardware can examine only a fixed portion of a basic block during any given decode cycle and, without software scheduling, easily misses opportunities for parallel execution. Even with out-of-order issue, software scheduling of basic blocks is desirable.

However, when scheduling instructions across basic-block boundaries, the advantage of software is not so clear. Figure 11-23 shows how the code of Figure 11-2 (which was used to illustrate trace scheduling) is executed on a processor with speculative, out-of-order issue. This code has software scheduling applied only at the basic-block level. Figure 11-24 shows how the loop of Figure 11-6 (which was used to illustrate software pipelining) is executed on the same processor, again with the

Decode				Execute					Writeback		Cycle
I0	I1	I2	I3	ALU1	ALU2	SHF	LS	BRN	R1	R2	
1:sll	2:add	3:load	4:add								1
5:store	6:cplt	7:jmpf	9:add			1:sll					2
10:load	11:add	8:add	12:store	2:add	9:add				1		3
					8:add		3:load		2	9	4
							10:load		8		5
					4:add				3		6
				6:cplt			5:store		4	10	7
				11:add				7:jmpf	6		8
							12:store		11		9

Figure 11-23. Code of Figure 11-2 with Basic-Block Scheduling and Out-of-Order Issue

I0	I1	I2	I3	ALU1	ALU2	SHF	LS	BRN	R1	R2	Cycle
load/1	fmul/1	add/1	clt/1								1
store/1	jmpt/1			add/1			load/1				2
load/2	fmul/2	add/2	clt/2		clt/1				x		3
store/2	jmpt/2			add/2		fmul/1	load/2	jmpt/1	x	x	4
load/3	fmul/3	add/3	clt/3		clt/2				x		5
store/3	jmpt/3			add/3		fmul/2	load/3	jmpt/2	x	x	6
load/4	fmul/4	add/4	clt/4		clt/3				x		7
store/4	jmpt/4			add/4		fmul/3	load/4	jmpt/3	x	x	8
load/5	fmul/5	add/5	clt/5		clt/4		store/1		x	x	9
store/5	jmpt/5			add/5		fmul/4	load/5	jmpt/4	x	x	10
load/6	fmul/6	add/6	clt/6		clt/5		store/2		x	x	11

Figure 11-24. Loop of Figure 11-6 with Basic-Block Scheduling and Out-of-Order Issue

same level of software scheduling. In these examples, out-of-order issue achieves the same performance as trace scheduling and software pipelining. The instruction window, with assistance from branch prediction, buffers instructions from more than one basic block regardless of the dynamic path taken through the code. Out-of-order issue, with assistance from register renaming, schedules instructions regardless of where they appear in the original program. Register renaming dynamically assigns new registers to different loop iterations and provides a mechanism for recovery (analogous to compensation code) when branch predictions are wrong.

We cannot generalize from these examples and conclude that software scheduling across branches and out-of-order issue are essentially the same thing. They are instead quite different, *partial* solutions to the problem of discovering instruction parallelism despite branches in the code. In many respects, hardware and software approaches to scheduling across branches are duals. The capabilities of software and hardware are similar, but they have very different sets of advantages and disadvantages.

Hardware decides which instructions to issue based on dependencies and resource availability, and has no idea which instructions are in the critical path. Because hardware has no knowledge of the critical path, it can make bad decisions about which instructions to issue, actually increasing the execution time by allocating resources to instructions not on the critical path. The increase in execution time can be especially severe if hardware adds one or more cycles to the iteration time of a short loop that is executed many times. Software scheduling helps by allowing the hardware to fetch the important instructions first, making it more likely that hardware will execute these instructions at an appropriate time. Even with speculative, out-of-order issue, software can improve performance via scheduling and global code motion. Still, hardware does not know about the critical path and can make decisions that increase the execution time.

On the other hand, out-of-order issue is not the root cause of nonoptimum execution in hardware. The fundamental problem is that the interface between software and hardware does not provide for software to communicate to hardware information about the critical path. Out-of-order issue simply gives the hardware more leeway to make bad decisions. We have seen several examples in this chapter where, even with in-order issue, the execution schedule is not the same as the software schedule. The processor used in these examples still makes decisions about when to issue instructions and how to allocate resources, and can still make bad decisions, although we saw no examples where this reduced performance. The real solution to this problem is to give software complete control over the processor. The instruction set of a VLIW processor, an architecture originally conceived for software scheduling, does exactly this and is one of several possible approaches. However, any instruction set that allows hardware to be aware of the critical path is not going to be compatible with any existing microprocessor. Even if such an instruction set were to gain acceptance, it would be difficult to maintain software compatibility over multiple implementations of the processor architecture, because the code would be implementation dependent: the critical path always depends on the hardware.

For software to achieve the same performance as out-of-order issue, software must be able to move code across branches. With conventional instruction sets and applications, this leads to expensive compensation code which reduces performance. Worse yet, this code motion can cause the program to behave incorrectly in ways that cannot be corrected by compensation code, leading to conservatism in the scheduler which reduces performance even further. Hardware can have the same effect as code motion across branches in a way that is invisible to the rest of the system. Out-of-order issue allows the design of a microprocessor that–from an external point of view–fits the conventional notions of how a sequential processor should operate.

Most of the performance of a superscalar processor results from executing independent instructions without regard for intervening branches. Neither a hardware nor a software approach to executing instructions across branches is particularly simple. The choice between hardware and software is determined mainly by the costs and difficulties of out-of-order issue, the desire for software compatibility, and the complexity of the software scheduler. The choice is a matter of perspective and system requirements. The following chapter examines these trade-offs in the broader context of microprocessor evolution.

Chapter 12

Evaluating Alternatives:
A Perspective on Superscalar Microprocessors

I decided to write this book for a couple of reasons. One reason is simply that I was attracted by the opportunity to describe an interesting area of computer architecture from my own perspective. Exploring and evaluating new ideas–or old ideas from new perspectives–is useful in its own right. However, I also wanted to develop a structure for these ideas, ways of thinking about superscalar processors and the myriad trade-offs they present. While performing the research for this book, I was exposed to a wide range of opinions about superscalar processors and formed and discarded many of my own opinions. I wanted to document this experience, at least indirectly. If superscalar processors are important, as I believe they are, this will help readers judge the many ideas, assertions, and controversies that are sure to accompany the emergence of superscalar designs.

As with most technologies, microprocessors are not developed for their own sake. Microprocessor development is driven by user requirements and constrained by the costs that users and developers are willing and able to bear. A microprocessor will not be used if it does not meet users' requirements, if it is not readily available, or if it is too expensive in comparison to other alternatives. A microprocessor will not be developed unless there are a sufficient number of potential users willing to pay a sufficient amount for the technology, and the perceived risks are low enough that there appears to be a good chance for successful development and manufacture. There often are many alternatives available to users and developers–and value, costs, and risks often are difficult to determine–greatly complicating the trade-offs and generating a lot of controversy.

The development of RISC processors during the 1980s serves as a good illustration. RISC techniques helped satisfy the needs of performance-oriented users.

However, RISC performance came at the expense of binary incompatibility with existing microprocessors, lower instruction encoding density, and higher instruction-bandwidth requirements. The costs to users were made palatable, in a sufficient number of applications, because advancements in memory density and speed permitted less efficient instruction encoding and provided increased instruction bandwidth. The increased density of static memories permitted the design of large caches, and high-pin-count packages permitted the design of advanced, high-bandwidth memory interfaces. Also, compiler technology reduced the costs and inefficiencies associated with high-level languages, so that source-code compatibility could be maintained to a sufficient degree even though binary compatibility was sacrificed.

Yet, the advantages of RISC did not entice all users to abandon older architectures. Some applications simply did not require additional performance. For many applications, the requirement for performance was not strong enough to override the requirement for binary code compatibility. Moreover, the performance of the older architectures improved at an adequate rate for many users. At the end of the decade, though RISC techniques had gained widespread respect, the controversy between those willing and able to change to RISC and those unwilling or unable to change had not subsided noticeably.

What, then, are the prospects for superscalar microprocessors? The controversy generated by RISC processors is likely to be small in comparison to the controversy generated by superscalar processors. Superscalar techniques are certainly not the only techniques available to increase performance; other techniques might be more attractive. Performance is always important, but the performance advantage of superscalar processors is limited and is well below the performance advantage of RISC processors over CISC processors. Even a simple superscalar processor is more complex than a RISC processor, and is not much faster if it is too simple. Achieving performance with a superscalar processor requires a complex software scheduler, complex hardware, or both.

In this chapter, we will examine the prospects for superscalar microprocessors. We will explore alternative superscalar designs—and alternatives *to* superscalar designs—in light of the value and costs to users and suppliers. These economic criteria are the best, perhaps only, criteria to employ to judge between the alternatives. These criteria certainly *will be* employed by users and suppliers, independent of any notion of technical superiority.

For our purposes, the value of a technical alternative is determined primarily by its performance. Throughout this book, we have focused on general-purpose applications that require performance, but these applications are not appropriate for many of the techniques used in high-performance scientific computing. Given this, we have few remaining alternatives. Because of the nature of general-purpose applications, performance can be improved either by increasing the operating frequency or increasing the instruction-execution rate via superscalar or related techniques.

The cost of an alternative is determined by many factors. Cost trade-offs do not universally apply, so we cannot reach a single, definitive conclusion that applies to all microprocessors over all time for all users. However, we can develop ways of viewing the alternatives. It is my hope that this helps readers understand how different techniques can be appropriate in different situations.

12.1   THE CASE FOR SOFTWARE SOLUTIONS

RISC processors have value to users who require performance more than code compatibility, and CISC processors have value to users who require code compatibility more than performance. Because RISC users are more performance oriented than CISC users, there is more impetus for RISC designs to develop superscalar implementations than there is for CISC designs. Also, because RISC processors are used in applications where binary code compatibility is not of paramount importance, new superscalar implementations may not have to be code compatible with existing RISC implementations. By opening up the possibility of new instruction sets, we open up the possibility of designing the instruction set to match closely the requirements of a superscalar processor.

RISC techniques focus on defining an instruction set so that the hardware is simple and so that the combined operation of hardware and software is efficient. In an analogous fashion, an instruction set tailored to a superscalar processor can simplify the design of the hardware and permit efficient operation. Simplicity and efficiency are related, though not identical. Both have value to users and suppliers. In this section we will consider new instruction sets for simplifying the superscalar hardware and improving performance. We cannot examine instruction sets in detail, and will only briefly consider a few ideas. Following this, we will discuss the risks associated with a new instruction set.

12.1.1   Instruction Formats to Simplify Hardware

When hardware is responsible for detecting parallelism between instructions, the decoder is complex. Each decoder position must be able to decode any instruction in the instruction set, and the decoder must determine the data dependencies and resource conflicts between these instructions. Complicated routing controls are required, because decoded instructions and operands must be routed from any decoder position to any functional unit.

An instruction for a VLIW processor is easier for hardware to decode, because functional units and operands are controlled by fixed instruction fields. At each field, the decoder need only be concerned about the class of instruction that can appear in that field. The operands for this instruction are specified within the same field, so routing is not required. Unfortunately, this fixed format is extremely inefficient in the applications or the portions of applications that have little instruction parallelism. For example, consider a VLIW format where the instruction is capable of issuing five operations at the same time. In some sections of code, dependencies can restrict the issue rate to one operation per instruction. In these sections of code, only one-fifth of the decoder and instruction-memory space is used. The VLIW instruction encoding is quite inefficient except for programs with a lot of instruction parallelism.

The encoding efficiency of a VLIW instruction can be improved by a hybrid serial/parallel instruction set, as shown in Figure 12-1. This instruction set has two or more different instruction formats. The simplest format issues a single operation, and other formats issue various numbers of operations in parallel. Figure 12-1 shows two parallel formats: one for issuing two operations and another for issuing four operations. With the hybrid instruction set, the scheduler uses serial instructions in sections of code with little instruction parallelism, and parallel instructions in sections with more parallelism (the width of the parallel instruction may depend on the degree of parallelism). Because the instruction can be more closely matched to the amount of instruction parallelism, the encoding is more efficient than in a VLIW processor. Cohn and colleagues [1989] report a significant increase in code density using this approach, though their conclusions are for scientific applications.

The hardware decoder for this hybrid format is more complex than a VLIW decoder, but not as complex as a decoder that detects parallelism in a stream of scalar instructions. One of the decoder positions must recognize all instructions, but other decoder positions need only recognize subsets of operations. The instruction fetcher also must align instructions, because instructions are not all of the same width. However, the parallelism is explicit in the instruction, so the decoder is not concerned with dependencies, functional-unit conflicts, or complex routing. Intel's i860™ processor [Intel 1989a] uses this concept, having a single-instruction mode for issuing single operations and a dual-instruction mode for issuing an integer and a floating-point instruction in parallel. Intel's 80960CA processor [Intel 1989b] uses a re-

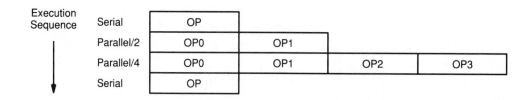

Figure 12-1. Hybrid Serial/Parallel Instruction Format

lated concept. The 80960CA does not have special parallel instruction formats, but recognizes parallelism only when certain classes of instructions are in certain decoder positions, with similar effect.

Another approach to instruction design addresses the problem of reducing the number of register ports required to provide the operands for parallel instructions. In Section 4.4, we observed that a four-instruction decoder requires only four register-file ports. Figure 12-2 illustrates an instruction format that facilitates restricting the number of registers accessed by a four-instruction decoder, doing away with the complex register-port arbitration. The instructions which occupy a single decode stage are grouped together, with a separate *register access* field. The register access field specifies the register identifiers for four source operands and four destination registers. Each destination-register identifier corresponds, by position, to an instruction in the decoder, and instructions do not need to identify destination registers. As Figure 12-3 shows, each instruction identifies source operands by selecting among the source- and destination-register identifiers in the register access field. Identifying the destination register of a previous instruction as a source indicates a dependency on the corresponding result value. For example, in Figure 12-3, the second instruction depends on the result of the first instruction, because the field *D0* specifies a source operand of the second instruction.

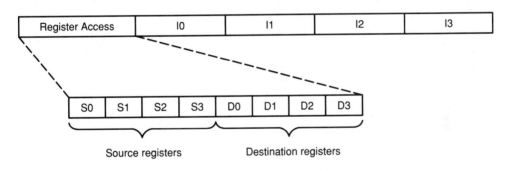

Figure 12-2. Parallel Instruction Format for Limiting the Number of Register Ports

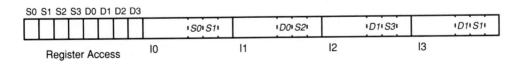

Figure 12-3. Example Operand Encoding Using Instruction Format of Figure 12-2

With this approach, operands are easily accessed during the decode stage. The source-register identifiers in the register access field are applied to the register file, and each instruction simply selects operands from among the accessed operands. If a group of instructions requires too many registers, software can insert *no-ops* in the instruction sequence to reduce the register usage of the group. Of course, each decoder position has to recognize every instruction and perform operand routing. However, the operand routing is simplified by the instruction encoding, and the instruction explicitly identifies data dependencies. This approach also allows several instructions within an instruction group to access the same operand register without using additional register-file ports. This sharing is difficult to accomplish in hardware: for example, a four-instruction decoder requires 28 comparators for matching source-register identifiers, and register-port sharing complicates the routing controls.

To take advantage of this instruction format, the software scheduler groups instructions by decoder boundaries and arranges the register usage of these instructions so that the group of instructions accesses no more than four total register operands. Meeting this restriction may require the insertion of *no-op* instructions into the group. This technique also may involve padding an instruction group so that it does not straddle branch targets, because an instruction after a branch target cannot indicate a dependency on an instruction that lies before the branch target. The instruction before the target may or may not be executed, depending on the path through the code, and so the dependency may or may not hold.

12.1.2 Instruction Formats for Scheduling Across Branches

The ideas presented in the previous section dealt with simplifying the hardware, but simplifying the hardware has only a weak performance benefit. The principal advantage of simple hardware is that the hardware design time is reduced, but a design that focuses just on simplifying the hardware runs a risk of being less preferred by users, compared to a design that has both simple hardware and increased performance.

We have seen that most instruction parallelism spans branch boundaries. Chapter 11 discussed at length the difficulties caused by software scheduling across branches. The biggest problem is caused by instructions that are executed when they should not be executed, because a branch outcome is different than the outcome predicted by software. This section considers hardware for undoing the effects of incorrectly executed instructions. This hardware allows a software scheduler to increase performance by scheduling instructions across branches, without the associated problems and inefficiencies and without the expense of out-of-order issue.

Figure 12-4 shows an instruction format that supports software code motion through branches. This instruction format and the following discussion are based on the TORCH proposal [Smith et al. 1990]. The instruction format shown in

Op Code	B	source1	source2	destination

Figure 12-4. Instruction Format to Support Code Motion Across Branch Boundaries

Figure 12-4 is a straightforward RISC format, except for the *B* bit. Unless the scheduler can move an instruction so that compensation is performed merely by copying the instruction to an alternate path, the scheduler can move an instruction only to the preceding basic block, for reasons described later. When the software scheduler moves an instruction up through a branch in an operation called *boosting*, the *B* bit is set. An instruction that is not moved from its original basic block has a *B* bit of zero. The *B* bit allows the hardware to identify boosted instructions and permits the hardware to cancel the effects of incorrect execution without the need for compensation code.

If a decoded instruction has the *B* bit set, the hardware places its result value into a *shadow register file* when the instruction completes (see Figure 12-5), and marks the corresponding entry as active. Each entry in the shadow register file corresponds to an entry in the register file. After the result is written into the shadow register file, subsequent instructions can refer to this value; these instructions would have followed the boosted instruction in the original code, and their results also are written into the shadow register file. In general, an instruction can refer to values either in the register file or the shadow register file, but the hardware controls which register file receives the result values. When a decoded instruction has a zero *B* bit, its result value goes directly to the register file.

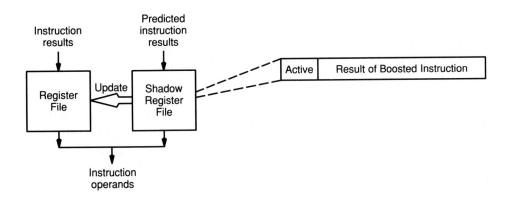

Figure 12-5. Shadow Register File

The shadow register file holds instruction results until the results can be correctly placed into the register file. Each time that a branch has the outcome predicted by software (a bit in the branch instruction indicates the expected outcome), each active entry in the shadow register file has had its intervening branch successfully predicted. The hardware writes each active entry in the shadow register file into the corresponding entry of the register file. All writes occur in a single cycle. Each time that a branch instruction does not have the outcome predicted by software, the hardware discards all entries in the shadow register file.

The shadow register file defers the updating of processor state by boosted instructions until it is certain that these instructions should have been executed (there is a structure similar to the shadow register file for stores to the data memory, which we are ignoring in this discussion). This eliminates the need for compensation code. Since the scheduler does not move instructions down through branch targets (see Figure 11-4a, page 208), there is no need to compensate for these instructions by avoiding their execution off the trace, and the original control flow of the program is preserved. All compensation can be done by hardware discarding the state updates of boosted instructions or by software copying boosted instructions off the trace.

Smith and colleagues [1990] report that this approach, with software scheduling, achieves roughly the same performance as out-of-order issue, and the hardware is much simpler. Note that the shadow register file and the register file can be constructed as a single memory cell, reducing the interconnection area required to update the register file from the shadow register file in a single cycle.

12.1.3 The Costs and Risks of Software Solutions

An instruction set that has been designed exclusively for a superscalar processor is not compatible with the instruction set of any scalar processor. The question of whether or not code compatibility is important is the object of much heated debate, but the question is best answered on the basis of the other evaluations in this chapter. Simply put, a user is willing to change instruction sets if the perceived value of the change is worth the costs of changing. A developer is willing to change if it lowers development and manufacturing costs, maintains or increases the number of users, and has acceptable risks.

The value and costs of changing instructions sets must be determined for each processor application. For example, in an engineering workstation, the value of performance is high, and, because of high-level languages, the cost of changing the instruction set is low. In a personal computer, on the other hand, the value of performance is lower, and, because the option of recompiling is not available, the cost of changing the instruction set is much higher. As we have already noted, RISC processors are used primarily in applications where code compatibility is not of paramount importance. The question is whether or not a superscalar processor offers a sufficient performance advantage to convince users–who have already made a significant investment converting to a RISC instruction set–to convert once more.

If it were true that the superscalar instruction set provided benefits over a RISC instruction set comparable to the benefits of a RISC instruction set over a CISC instruction set, the trade-off would be clearer. RISC processors were able to attract a sufficient number of users by offering about a factor-of-three performance benefit over CISC processors. In my experience, this performance benefit was just barely enough for many users. Superscalar processors appear to offer a factor-of-two performance benefit or less, reducing the value of a new, superscalar instruction set and increasing the risk that it will not be adopted. Still, the costs of converting to a new instruction set depend on the application, potentially making the performance benefit of the conversion attractive in some cases. So it is useful to examine both sides of this question.

Certainly, some users will change instruction sets for almost any performance improvement. Some applications place such a high value on performance that many alternatives are possible. For example, logic verification of a complex chip can consume weeks or months of processor time. In such an application, performance improvements have a very high value, and it is relatively easy to justify all sorts of costs.

When a processor is intended for applications that place a very high value on performance, the primary risks are presented by competitive alternatives. If there are no other alternatives, changing instructions sets for performance clearly is acceptable. However, the fact that performance has high value, itself, often opens up many other alternatives. Alternatives such as a design in a high-speed technology or a design that uses hardware-only superscalar techniques might achieve equivalent or better performance without incompatible software. If these other alternatives cost less than the cost of incompatible software, they will be preferred. Thus, the risks of a software-incompatible design depend on the likelihood of competitive alternatives, and on the costs of software incompatibility.

The likelihood of competitive alternatives to a software-incompatible design is affected by the relative difficulties of the alternatives. Improvements in technology can alter the feasibility of the various alternatives available to users over the course of development, changing the nature and characteristics of the alternatives over time. For example, a software-intensive approach requires both a software scheduler and the processor hardware to offer performance. The major risk is that two developments, one for the hardware and one for the software scheduler, must be successful and complete at the same time. If either of these elements is delayed, another alternative–one that may not have been originally anticipated–may become available and be used instead.

In some applications, the processor is a captive design: it is designed only as part of a larger project, and there is no competition. Also, in some applications, the processor is dedicated to executing a single software program, and there is little cost to implement this program in a new instruction set. Such special-purpose, application-specific processors should be designed to have simple hardware, because often few of these processors are built. If few processors are built, the hardware development cost must be very low so that the cost per processor is low. This argues for a standard implementation technology such as gate arrays or discrete logic. In applications of this type, hardware simplicity is much more important than software compatibility, because implementation alternatives are limited.

Though software incompatibility may not present much of a problem for an application-specific processor, the situation is quite different with a general-purpose processor in widespread use. In this case, software incompatibility causes many difficulties. A processor architecture proliferates by addressing the needs of cost-sensitive users: the lower the costs, the larger the number of users. Also, a processor in widespread use executes very many different programs, so that the costs of changing the instruction set are very high. Here, a new instruction set has a double disadvantage: the inclination of most users is to resist any cost, but the cost of changing the instruction set is high. It is nearly certain that the performance of a superscalar processor is not high enough to justify the costs of changing the instruction set. The superscalar processor must be software compatible with previous implementations.

How can software compatibility be so important? After all, haven't a number of users already demonstrated that they were able to convert to new, RISC architectures? The response to this question is that the implications of software incompatibility for superscalar processors are simply different from the implications for RISC processors. Many users converted software to a RISC architecture in anticipation that they would not have to convert more than once, and that they could amortize this cost over several generations of RISC processors. An incompatible superscalar processor requires that users incur this expense once again. Worse yet, if the software depends on a particular hardware implementation–as it does, for example, when software insures that data dependencies are satisfied–this cost must be incurred every time a new hardware implementation has different operation latencies than the previous implementation. There also must be a different version of the scheduler (or configuration of the scheduler) for each different processor version.

We have seen also that some high-level languages, such as ANSI C, do not allow extensive code reorganization. Extensive software scheduling would render some programs incompatible even at the source-code level. This is unlikely to be much of a problem for existing applications that are simply recompiled. It is more of a problem in the development of new applications, where interactions via a debugger expose a user to the code motions performed by the scheduler. The point here is not that this problem is insurmountable, but rather that some standard programming environments do not anticipate the extensive code motion required for performance in a superscalar processor, particularly code motion across branch boundaries.

To summarize, it is not out of the question for a superscalar processor to have a new, incompatible instruction set. In some special-purpose applications, software incompatibility yields an overall benefit because the hardware is simple. However, in widespread, general-purpose applications, software incompatibility presents many risks to developers. It requires dual developments, one for the scheduler and one for the processor hardware, and requests that users regenerate software. Each time users are requested to regenerate software, they are likely to evaluate other alternatives. The risks of being late in developing both the compiler and hardware, coupled with the low performance potential of a superscalar processor, makes it possible–even probable–that users will find other alternatives that are more acceptable.

12.2 THE CASE FOR HARDWARE SOLUTIONS

Hardware-oriented superscalar techniques make it possible to improve performance while preserving, in appearance, the sequential behavior of an existing scalar architecture. This is the sole value of hardware solutions. A new superscalar implementation can be made software compatible with a previous scalar implementation, and users are not required to change their concept of how the processor behaves. However, we cannot accept this benefit in isolation, any more than we could accept, for software solutions, the benefit of hardware simplicity. Where software solutions have the cost of incompatibility, hardware solutions have the cost of additional hardware. Where the tandem development of both a software scheduler and hardware is risky, the development of complex hardware also is risky.

A hardware-oriented superscalar design must compete with other hardware-oriented means to improve performance. The easiest way to increase the performance of a processor, for both users and developers, is to increase the processor's operating frequency. This simply requires improvements in the implementation technology. The investment in technology needed to increase the frequency is made for many components and products, not just microprocessors. Since the costs are amortized across many products, the improvement in processor frequency is almost free. This approach requires little architectural design and no software changes.

The principal impediment to increasing the processor's operating frequency is that, for a given packaging technology, the speed of the processor's memory interface and memory system do not improve at the same rate as the processor's operating frequency. The solution to this problem is obvious, however. Along with the frequency improvements, technology advancements allow higher memory densities which can be used to implement larger and larger caches in proximity to the processor. The design of a cache is relatively easy, so development costs are not increased very much. From a user's perspective, larger and larger caches have approximately a fixed cost over time, owing to technology improvements, but larger caches provide higher and higher performance over time, so the cost is justified.

The ability to improve the performance of a scalar processor through simple frequency scaling has a very important bearing on the risks of developing a superscalar processor. We will first examine two different models of frequency scaling–one that applies in performance-oriented applications and another that applies in cost-sensitive applications–then examine the attractiveness of a superscalar design in each case.

12.2.1 Two Models of Performance Growth

The nature of semiconductor and packaging technology tends to segment users into two groups: performance-oriented users who need performance at almost any cost and cost-sensitive users who need the best performance they can get at a given cost. Because semiconductor economies depend very heavily on having a large number of users, and because there are many more cost-sensitive users than performance-oriented users, cost-sensitive users tend to drive the development of mainstream semiconductor technology. However, performance-oriented users still get attention because of their willingness to pay–they just remain out of the mainstream where costs are always high [Jouppi 1989a]. The performance supplied to both groups increases over time, but performance growth is constrained in different ways.

To illustrate this point, consider two example implementations of a single architecture. The first is for performance-oriented users: a multiple-chip implementation using emitter-coupled logic (ECL) technology. The second is for cost-sensitive users: a single-chip implementation using complementary metal-oxide semiconductor (CMOS) technology. The ECL processor requires the expense of a multiple-chip implementation for several reasons, but one reason is that there is not much motivation for semiconductor developers to improve the circuit density of ECL technology (that is, the number of circuits on a single chip), because there are not many users compared to the number of users of CMOS technology. The performance growth of the ECL implementation is determined in large part by the packaging and wiring technologies that connect these chips. On the other hand, the CMOS processor fits on a single chip, because, for cost reasons, the most important design objective is that the processor be implemented on a single chip. The performance growth of the CMOS implementation is determined in large part by what can cost-effectively be fit on a chip.

For our purposes, we assume that performance-oriented and cost-sensitive users scale frequency differently. This provides two different scenarios for examining the value of a superscalar design. We assume that, as the frequency is increased and the cache size is increased, the performance-oriented user reduces the cache reload time correspondingly. This may require new memory technology, new packaging technology, and other high-cost items, but the performance-oriented user is willing to pay. In contrast, the cost-sensitive user is not able to reduce the reload time significantly. The speed of commodity memories does not increase as fast as the processor operating frequency, and the performance of the memory interconnection is constrained by the inductance, capacitance, wire-propagation delays, and clock skews determined by the packaging technology. In the cost-sensitive scenario, both users and developers would resist adopting new packaging technologies, particularly because of the capital costs involved in tooling equipment. These cases are extreme, but we are interested in examining design boundary points.

Because the relative merits of a superscalar design can depend on the processor configuration, we also consider potential designs over five generations. The first generation has small, 2-Kbyte caches, and the cache size doubles with each generation. The frequency increases by 60% in each generation, from a normalized frequency of unity for the first generation. These parameters are by no means the only ones that could be chosen. There are an infinite number of possibilities, but we are only trying to gain insight into the merits of a superscalar design, not to do an exhaustive study. The selected parameters simply allow the examination of a reasonable number of generations while avoiding ridiculous assumptions for either the first or last generation.

Figure 12-6 shows the speedup of each generation using the performance-oriented scenario. Here, the reload time is held constant at 12 cycles (this is to keep the range of actual reload times within reason). The "simple" superscalar design in Figure 12-6 has a two-instruction decoder and uses in-order issue. The "complex" superscalar design has a four-instruction decoder, out-of-order issue, branch predic-

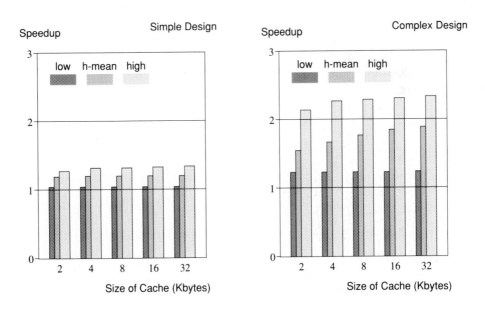

Figure 12-6. Speedup as a Function of Cache Size–Scalable Reload Time

tion, and register renaming. Both the simple and complex processors have two ALUs and allow parallel execution of any instructions that use different functional units (rather than, say, just parallel execution between floating-point and integer operations as with some existing superscalar designs [Apollo 1988]). The relative performance of the complex design increases with each generation, because this design takes greater advantage of cache hits than the scalar processor.

Figure 12-7 shows the speedups, again of simple and complex designs, of each generation using the cost-sensitive scenario. Here, the reload time is the same for each generation and is equivalent to 4 cycles in the first generation, growing to 31 cycles in the last generation (again, to keep the range of actual reload times within reason). The performance of the complex design is relatively flat, because the cache-miss time represents an increasing proportion of the execution time. Performance of both the scalar and superscalar processors tends toward saturation, limited by memory performance.

Plotting the speedups of superscalar processors, however, does not expose their risks. These plots assume that the scalar and superscalar designs cost the same, are available at the same time, and are otherwise equivalent except that one happens to be superscalar. We must evaluate the possible impacts of hardware complexity. The superscalar design might have a longer cycle time, or might be available later because it takes longer to design. The next two sections consider both of these risks, for both performance-oriented and cost-sensitive designs.

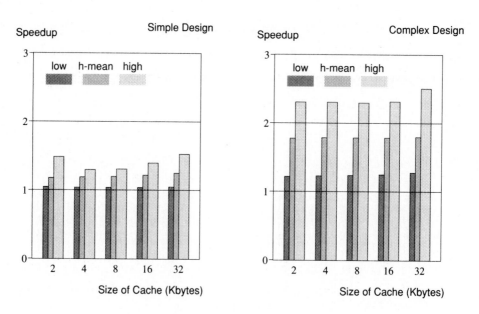

Figure 12-7. Speedup as a Function of Cache Size–Constant Reload Time

12.2.2 Estimating Risks in a Performance-Oriented Design

We are assuming that, in a performance-oriented design, the cache reload time of each generation is reduced in proportion to the reduction in processor cycle time. Because the cache size increases also, the reload penalty (the product of the reload time and the miss ratio) decreases with each generation. This is the most optimistic assumption in favor of simply scaling the frequency to improve performance. The decreasing reload penalty causes the improvement in performance to be larger than the improvement in frequency.

Figure 12-8 plots the performance growth of several generations of scalar and superscalar processors under these assumptions. These plots are derived from simulation, using the harmonic mean of performance over all benchmarks. The first plot in Figure 12-8 shows the performance growth, over five generations, of a scalar processor, compared to the performance growth of the simple and complex superscalar processors defined in the preceding section. All processors have equivalent-sized caches. The performance of each generation is shown relative to the performance of the scalar processor in the first generation.

The second plot in Figure 12-8 compares the performance growth of the same three processors, except that the superscalar processors have caches that are one-half

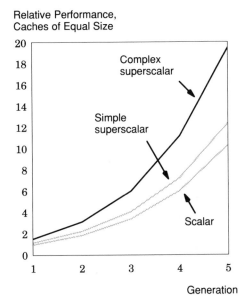 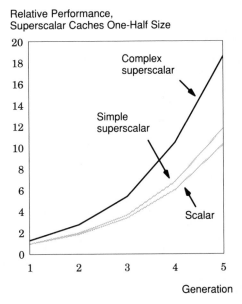

Figure 12-8. Performance Growth of Scalar and Superscalar Processors Relative to First Scalar Generation–Scaled Reload Time

the size of the caches of the scalar processor. This accounts, roughly, for the possibility that the superscalar hardware may consume part of the chip area that is allocated to caches in the scalar processor. The first plot in Figure 12-8 should be viewed as comparing scalar and superscalar processors where the superscalar processors are more expensive, and the second plot should be viewed as comparing processors that cost about the same. This view is more accurate for the first three generations than for later generations, because the superscalar hardware does not grow very much over time. Superscalar hardware is unlikely to cost very much, relative to the cost of caches, as the technology improves and the caches get very large. However, these rough plots are sufficient for our analysis.

By themselves, the plots in Figure 12-8 do not provide very much information except for the unsurprising revelation that the superscalar processors are always faster than the scalar processors over many generations. Our purpose in presenting these plots is to explore the risks that accompany the performance advantages of the superscalar processors. There are two ways to gauge the magnitude of these risks (see Figure 12-9). The cycle time of the superscalar processor may be longer than the cycle time of the scalar processor, reducing or eliminating the performance advantage of the superscalar processor. In addition, the design time of the superscalar processor may be longer than the design time of the scalar processor, meaning that the scalar processor has a chance to catch up via frequency scaling. The risks of the

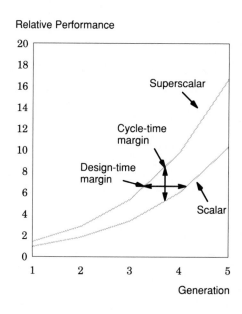

Figure 12-9. Cycle-Time and Design-Time Margins of Superscalar Processor

superscalar processor depend on the likelihood of either of these possibilities. The likelihood in turns depends on by how much the superscalar cycle time or design time must be increased before the performance of the scalar and superscalar processors are equivalent, in which case there is no advantage to the superscalar processor.

We define *cycle-time margin* to be the increase in cycle time, as a percentage of the scalar cycle time, that must be incurred in the superscalar processor for its performance to be equivalent to the performance of the scalar processor. We define *design-time margin* to be the increase in design time, as a percentage of the scalar design time, that must be incurred in the superscalar processor for its performance to be equivalent to the performance of the scalar processor (assuming that the scalar frequency can be increased continuously). Of course, the design of a superscalar processor should have the goal of equivalent cycle time and availability compared to the scalar processor. The cycle-time margin and design-time margin simply give insight into the consequences if these goals are not met.

Figure 12-10 plots the cycle-time margins of the superscalar processors, derived from the data in Figure 12-8, and Figure 12-11 plots the design-time margins. The superscalar processors clearly have more risk (indicated by lower margins) in earlier generations than in later generations, particularly if the caches are reduced to accommodate the superscalar hardware. The margins of the complex design stead-

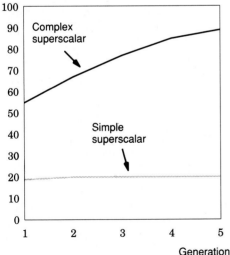
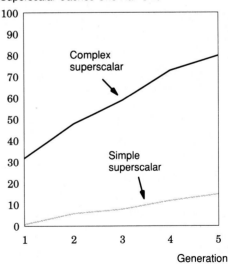

Figure 12-10. Cycle-Time Margin of Superscalar Processors by Generation–
Scaled Reload Time

Percentage of Scalar Generation Design Time, Caches of Equal Size

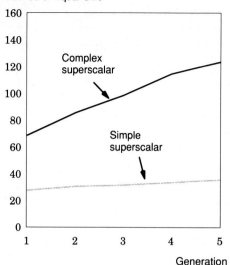

Percentage of Scalar Generation Design Time, Superscalar Caches One-Half Size

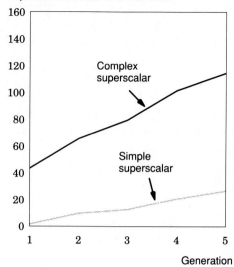

Figure 12-11. Design-Time Margin of Superscalar Processors by Generation–
Scaled Reload Time

ily increase with each generation, and the risks decrease, because this design takes more advantage of larger caches than the scalar processor. The value of a cache hit is higher in the complex superscalar processor. In contrast, the margins of the simple design do not change very much over successive generations. Of course, we would expect that the simple design would not require much margin–which is fortunate, because there is not much margin.

The design-time margin of the complex superscalar processor in the final generation is particularly interesting. Even if the superscalar processor has half-sized caches, it is about a generation ahead of the scalar processor if available at the same time. Viewed another way, the superscalar processor has an opportunity to get about a generation ahead of the scalar processor, between the fourth and the fifth generations, without doubling the cache size from 16 Kbytes to 32 Kbytes. At this point in technology development, the size of the superscalar hardware is likely much less than the size of the caches. The raw cost of the superscalar hardware is lower than the cost of the caches, yet the superscalar hardware provides considerably more performance.

12.2.3 Estimating Risks in a Cost-Sensitive Design

We are assuming that, in a cost-oriented design, the reload time of each generation is constant, because users are not willing to pay for the packaging, interconnection, and memory technology required to scale the reload time. Expressed in terms of the number of processor cycles required to perform a reload, the reload time actually increases with each generation. Because the cache size increases also, the miss ratio decreases, and the reload penalty does not increase at the same rate as does the reload time, though the reload penalty still is worse with each generation. This is the most pessimistic assumption against simply scaling the frequency to improve performance.

Figure 12-12 plots the performance growth of several generations of scalar and superscalar processors under these assumptions. These plots correspond to the plots in Figure 12-8 and have the same scale to aid comparisons between the figures. Not surprisingly, performance does not grow nearly as fast when the reload time is not scaled, because memory performance tends to dominate overall performance.

Figure 12-13 plots the cycle-time margins of the superscalar processors, for comparison with Figure 12-10, and Figure 12-14 plots the design-time margins, for comparison with Figure 12-11. This risk assessment is quite different from that in the performance-oriented scenario. If the superscalar caches are the same size as the scalar caches, the superscalar processors have nearly constant margins, and the risk is about the same over all generations. On the other hand, if the superscalar processors have half-sized caches, the margins decrease over time; the reason for this is that

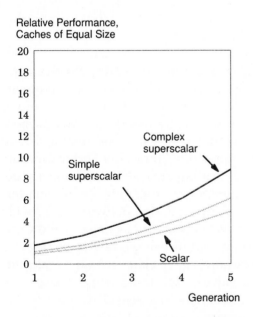

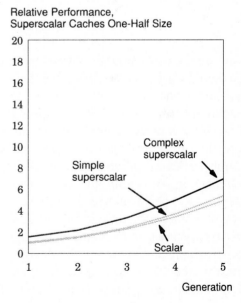

Figure 12-12. Performance Growth of Scalar and Superscalar Processors Relative to First Scalar Generation–Constant Reload Time

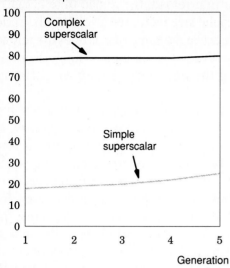

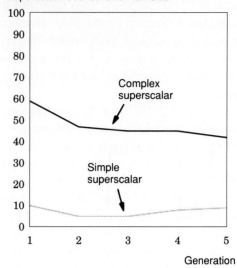

*Figure 12-13. Cycle-Time Margin of Superscalar Processors by Generation–
Constant Reload Time*

the smaller caches put the superscalar processors at an ever-increasing disadvantage in their ability to compensate for the increasing reload time. However, this problem is more academic than real, because this problem is most severe for later generations where it is less likely that the superscalar processors really require half-size caches.

Interestingly, superscalar techniques have the most value and lowest risk in cost-sensitive applications. Superscalar techniques allow processor performance to improve beyond what is possible with frequency scaling, for users who cannot afford changes in packaging and interconnection technology. In fact, by comparing the superscalar performance-growth curve for the complex design in Figure 12-12 with the scalar growth curve in Figure 12-8, we see that superscalar techniques allow the cost-sensitive user to experience roughly the same performance growth as the user of a performance-oriented scalar design. Furthermore, superscalar techniques take the best advantage of large-scale integration. The value of the most expensive processor component–the caches–is increased if the caches are augmented by superscalar hardware. From the user's standpoint, the memory system yields another cost savings: the memory costs decrease over time, because memory performance is constant as the memory technology improves.

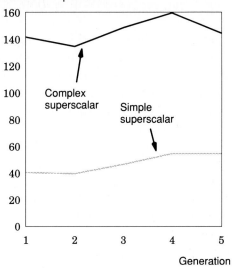

Percentage of Total Scalar Design Time, Caches of Equal Size

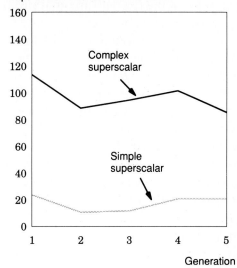

Percentage of Total Scalar Design Time, Superscalar Caches One-Half Size

Figure 12-14. Design-Time Margin of Superscalar Processors by Generation– Constant Reload Time

12.2.4 Putting Risks in Perspective

There are quite a few assumptions behind the observations in Sections 12.2.2 and 12.2.3. The following were all treated as given: operating-frequency growth, reload times, cache organization (direct mapped), cache-reload strategy, and superscalar features. We also used fixed operation latencies for each generation, ignoring the likelihood that increasing circuit densities can be used to shorten the latencies, especially for floating-point operations. Just varying a few parameters yielded a lot of data, however. Our goal is not to examine all alternatives exhaustively, but to develop a method for assessing risks so that we can avoid dismissing or endorsing superscalar techniques in a vacuum. This analysis also illustrates that the risks and attractiveness of a superscalar design can depend significantly on the assumptions underlying the design.

As with any abstract concepts, the cycle-time margin and design-time margin help provide insight but also hide information. For example, real frequency growth and development times are not nearly as smooth as we have implied. Also, it is hard to know what a "good" margin is. A margin of 20% is only 4 nanoseconds in a 50-megaHertz processor. The performance advantage of the superscalar processor is negated if only one path, out of millions, uses up the cycle-time margin. Furthermore, the risks are not independent: a design cannot sacrifice both the cycle-time margin and the design-time margin, because the resulting design will be both late and slow.

There are also other factors to consider in assessing the risks to cycle time and design time. The cycle time may be determined by factors that have nothing to do with superscalar features. For example, if the cycle time is dominated by off-chip paths, there may be little risk that the superscalar hardware will increase the cycle time. A similar observation applies to the design time. Often, the design time between generations is determined by the time it takes to develop the semiconductor technology, rather than the time it takes to develop the processor itself. If the time between processor generations is technology limited, the time to design the superscalar hardware may not increase the overall development time.

It can be difficult to determine the value of being ahead of the performance growth of a scalar processor. In some applications, user performance requirements may grow at about the same rate as scalar performance, or may grow in one generation and not grow at all in another generation (because costs are reduced in this other generation). If the performance supplied by a superscalar processor is ahead of users' requirements, users will not see much value in the superscalar design. In this case, there would be no reason to pay for the development of a superscalar processor.

Risks are incurred not only by undertaking a development, but also by not undertaking a development. In particular, nearly all developers can afford to add caches to a processor, so adding caches alone yields no particular competitive advantage. If a significant number of users need increased performance growth, there is an advantage to the developer that can successfully design and build a superscalar processor. From the user's perspective, the system hardware and software for a superscalar processor can be identical to those for a scalar, cache-based processor, except that the performance in each generation is better than the performance of a scalar processor. Superscalar techniques can have a long-term value to users that is in line with the long-term costs. Also, users may prefer a given scalar implementation partly because of the availability of a compatible, higher-performance, superscalar implementation, even if the higher-performance version is not used very much.

The feasibility of a superscalar design can change over time. This is especially true as performance becomes more and more limited by the main memory. At some point in development, it is likely that superscalar hardware is a better choice than increasing the size of the caches. The risks of designing a superscalar processor can be reduced by first designing a simple superscalar processor, then a more complex one in a later generation. Alternatively, the development of the superscalar processor can span two or more scalar generations, to be made available at the point where performance is expected to be memory limited.

Considering all factors, it is hard to imagine that superscalar techniques would not be attractive at any point in the lifetime of a microprocessor architecture.

Superscalar techniques can help performance growth for many processor generations, and they increase the efficiency of the hardware. They are attractive at some point because they permit performance to grow faster, and with less expense, than do new circuit technologies, package technologies, and new system approaches such as multiprocessing. Ultimately, superscalar techniques buy time to determine the next cost-effective techniques for increasing performance.

Appendix

A Superscalar 386

This book has concentrated primarily on RISC architectures, because super-scalar techniques can be difficult enough to understand without the added complexity of CISC instruction sets. This approach helped us explore the issues and trade-offs involved in the design of superscalar microprocessors, but leaves us with little specific information about the sorts of problems encountered in a superscalar CISC processor and what techniques might be employed to solve these problems.

In this appendix, we will address this shortcoming–at least to some extent–by examining a superscalar implementation of Intel's 386 architecture. Intel has stated that this architecture, first implemented as the 80386, is to be used as the basis of a number of future generations, of which the i486™ is the most recent example. I chose the 386 architecture because, of all CISC microprocessor architectures, it is the one most likely to have a superscalar implementation. Furthermore, the 386 is quite complex, providing a number of examples of the kinds of additional problems that must be solved in a superscalar CISC implementation. Unfortunately, this complexity also prevents us from exploring this subject in depth. Intel–and anyone else who might attempt this implementation–is going to have to expend a lot of resources on the effort, and we cannot hope to match the depth of analysis and study needed to chose the proper implementation techniques. However, we can gain a broad, general perspective of such a design.

A.1 THE ARCHITECTURE

It is necessary to spend only a few minutes studying the 386 architecture [Intel 1986] before one thing becomes very clear–the 386 was not defined with a superscalar implementation in mind. This statement should be obvious, and seems a little unfair. None of the popular RISC architectures developed in the mid-1980s was defined with a superscalar implementation in mind, either. Only IBM's RIOS architecture [Groves and Oehler 1989] and Intel's i860 architecture [Intel 1989a] included the concept of pipelines wider than one instruction in the original implementation. Yet, regardless of whether the architects during the original definition were specifically targeting a superscalar implementation, a superscalar RISC is not nearly as difficult as a superscalar CISC.

Why is this so? I think the reason is simply that most RISC architectures were defined with a pipelined implementation in mind and that this is sufficient to greatly simply the superscalar implementation. It does not matter that the superscalar implementation was not considered from the onset. The features of an architecture that make it easier to achieve parallel execution of instructions in different pipeline stages–such as having fixed-format, fixed-length instructions, explicit instruction operands, few side effects, a load/store architecture, and so on–also make it easier to achieve parallel execution of instructions in the same pipeline stage. There are only a few things that might be done differently. For example, delayed branches with a single delay instruction are quite helpful in a pipelined scalar processor but are pretty useless in a superscalar processor because they do not overcome enough of the branch-delay penalty. However, these differences are minor and do not have much effect on the basic complexity of the superscalar implementation.

In contrast to RISC architectures, CISC architectures were defined at a time when the principal implementation technique was microcode interpretation of the instruction set and when pipelining was considered to be an exotic technique. Design goals were oriented more toward deciding which operations should be combined into instructions than designing operations so that they could be overlapped. Because of microcode interpretation, almost anything could be done with the definition of the instruction set–and generally just about everything was done. Unfortunately, this makes it painful to implement a pipelined version of the architecture, and extremely painful to implement a superscalar version. In this section, we will look briefly at a few of the difficulties.

A.1.1 Instruction Format

Figure A-1 shows the instruction format of the 386. There are many variations on this format, and Figure A-1 shows the longest typical instruction (actually, an in-

| prefix | opcode | opcode | mod r/m | s-i-b | 0,1,2,or 4 bytes of displacement | 0,1,2,or 4 bytes of immediate data |

Figure A-1. Instruction Format of the 386

value of "10000" in the *mod r/m* field selects the *BX* and *SI* registers in the 16-bit mode and the *EAX* register in the 32-bit mode.

But dependency checking is more difficult than just trying to determine which registers are being accessed. The presence or absence of a dependency also is determined by the data width of an instruction, which in turn determines whether the register operands are 8, 16, or 32 bits wide. This by itself is not much of a problem, but the 8-bit registers overlap the 16- and 32-bit registers, and the 16-bit registers overlap the 32-bit registers. This overlapping introduces another level of complexity, because the fields used to specify the overlapping registers for some data widths are also used to specify nonoverlapping registers in other data widths. For example, a value of "11100" can specify the *ESP* register, which is independent of the *EAX* register, or the *AH* register, which overlaps the *EAX* register and thus creates a dependency.

A.1.3 Memory Accesses

The difficulties with the 386 register architecture are pretty much repeated in the memory architecture. Because of the shortage of registers, there are many more operations involving memory than in a RISC architecture, and the register-to-memory operations destroy the source operand in memory. There is not much data on the frequency of memory operations in the 386, but, in the 8086 predecessor of the 386, memory accesses for operands are over twice as frequent as memory accesses in a typical RISC processor [Adams and Zimmerman 1989]. This means that there is a much greater likelihood of dependencies between memory accesses than in a RISC processor and that the memory dataflow has a greater impact on performance.

But like register dependencies, memory dependencies are difficult to detect. The 386 provides an extensive set of addressing modes to compute an effective address, then applies two levels of address translation to the effective address (segmentation and paging) before the physical address is determined. For example, the double-indexed addressing mode takes three address operands as inputs and performs two adds (with a simple shift of one of the operands) to form the effective address. The effective address is then added to the segment base address in a segment register selected by the instruction. The segment register, and thus the base address, can be overridden by a prefix byte, so that address translation uses a different segment register than the one specified by the instruction. The value resulting from adding the effective address to the segment base address is compared with a segment-limit address in the segment register to insure that the address is within bounds. Finally, as an option, page translation is applied to the address to determine the final physical address. All of these operations must be applied before a memory dependency can be checked.

Even after the physical address has been determined, dependency checking is not trivial. The 386 allows unaligned memory access of data that is either 8, 16, or 32

---

bits wide. For example, an unaligned 32-bit value may span two 32-bit memory locations. The addresses of these two memory locations can be quite different, depending on the crossing of address boundaries. Hence, because of unaligned accesses, addresses must be compared by subtraction to determine whether or not there is a dependency, rather than a simple equal-to comparison, and the existence of the dependency is a complex function of the results of this subtraction and the data widths of the two accesses that might be dependent.

A.1.4 Complex Instructions

In addition to the basic complexity of the 386 architecture described in the foregoing sections, the 386 includes several very complex instructions. A few examples are:

- The *enter* (enter procedure) instruction pushes the run-time stack in external memory and copies multiple items of the calling procedure's stack to the called procedure's stack. The copy is intended to set up the *display* of the called procedure, supporting the scoping rules with nested procedures as in the Pascal programming language.

- The *pusha* (push all registers) moves all registers to the memory stack, and the *popa* (pop all registers) instruction loads all registers from the stack.

- The *cmps* (compare string operands) instructions compare two 8-, 16-, or 32-bit operands in memory and decrement two pointers to these memory operands.

- The *rep* (repeat) instruction appears before an instruction to indicate that the following instruction should be repeated a number of times. For example, the *rep* instruction can appear before a *cmps* instruction to compare two strings of arbitrary length. The *rep* instruction imposes on the following instruction a control flow commonly implemented by a compare, decrement, and branch sequence, causing the following instruction to have a very complex set of dependencies and resource utilization.

These instructions, and others like them, take a large number of input operands from either registers or memory, interact with a large portion of the processor resources, and possibly produce a large number of result values into either registers or memory. It would be extremely difficult to execute these sorts of operations in parallel with other instructions, and it very likely would not be worth the effort. Because there are so many operations with true dependencies within such instructions, they must be overlapped with a very large number of other, simpler instructions to take advantage of a superscalar implementation. However, because there is so much going on during the execution of these instructions, it is not likely that there is much else that the processor can find to do. Besides, even if the processor could achieve parallelism with these instructions, they are probably so infrequent that any perform-

---

ance benefit would be a very small portion of the total execution time. Since it is too difficult and not worthwhile to execute such instructions in a superscalar mode, a superscalar implementation probably would have two modes of execution: a slow, serial mode for the very complex instructions and a faster, superscalar mode for the simpler instructions.

A.2 THE IMPLEMENTATION

At this point, it is tempting to ask: Why would anyone go through the pain of implementing a superscalar 386? There are two very good reasons. First, because the 386 architecture is so important in personal-computer applications and because the number of processors that can be sold in this market is very large, there is enormous economic incentive to develop a superscalar 386. At the same time, the existing personal-computer market provides enough revenue to fund this expensive development. Second, the personal-computer market values code compatibility more than almost anything else. Indeed, the 386 itself carries a lot of baggage that is due only to compatibility with earlier Intel processors. But, on the positive side, the importance of compatibility gives the 386 architecture almost exclusive access to the personal-computer market. This yields a double advantage to the developer. The performance of the superscalar design is not as important as compatibility, so the design does not have to yield a tremendous performance improvement to be of value. Also, because applications for the 386 exclude the possibility of any processor that is not object-code compatible, the competition faced by the design is limited. Thus, it is important only that the superscalar 386 implementation be faster than the fastest implementation of the 386 architecture–for example, the i486–not that it be faster than the implementation of every other architecture. In any case, the risk of the development is fairly low.

The i486 implementation of the 386 architecture is a good example of these points. The 386 instruction set is quite difficult to pipeline, but the performance achieved by the i486 indicates that the designers employed aggressive pipelining. This is an illustration of what is possible given enough resources, motivation, and experience with an architecture.

In this section, we will give a cursory look at three techniques that might be employed in the design of a superscalar 386. All of these techniques assume that the implementation has some form of microcode–a pretty safe assumption, since there must be some provision for the most complex instructions. The first two techniques apply in a straightforward microcoded implementation, such as the 80386. These two techniques are somewhat analogous to the two techniques we have examined in this book: out-of-order issue and software scheduling. However, in the 386, both techniques must be implemented in hardware, because there is no option to perform software scheduling of existing applications. The third technique–the one I think is

---

most likely–is presaged by the i486. This technique is based on defining a "RISC core" of instructions that are able to take advantage of superscalar techniques.

A.2.1 Out-of-Order Microinstruction Issue

One way to approach the superscalar 386 design is to start with a pure microcoded scalar implementation, without the hardwired control and special techniques that appear to have been employed in the i486. The goal in this case would be to achieve parallel execution of microinstructions in a superscalar pipeline. Because the microinstruction sequence for a particular instruction probably contains many dependencies, it is important to allow out-of-order issue of microinstructions so that sequences from different instructions can overlap. This is the essence of the idea behind the HPS technique proposed by Patt and colleagues [Patt et al. 1985a] (the acronym HPS stands for *high performance substrate*). The advantage of this approach is that it converts complex instructions into sequences of microinstructions that are much easier to deal with. We would expect microinstructions to have many of the characteristics of RISC instructions, so that the techniques we have examined in this book would be easier to apply at the microcode level. Unfortunately, applying speculative, out-of-order issue at the microcode level is more difficult than it might initially appear.

One problem is simply the raw microcode bandwidth that is required. The microinstruction word of the 80386 is 37 bits wide. If we assume that about 2 to 3 microinstructions, on the average, are required to emulate a single instruction, then the typical instruction expands into approximately 80-100 bits of microinstruction. If we attempt, say, to execute two instructions per cycle, we must be prepared to deal with about 6 microinstructions per cycle. Consequently, we need quite a bit of hardware to uncover any significant amount of instruction parallelism.

The other difficulty with this approach is that it introduces an entirely new level of processor state and program control flow. One level is the architectural level, containing the program-visible state and the program branches. The second level is the microcode level, containing the microcode state and the microcode branches (microcode branches must be present to implement some of the complex instructions, such as *enter*).

The microcode state includes items such as microcode scratch registers, datapath registers, and so on. In general, a microinstruction can specify an item of state in either the program or the microcode level, and can specify this item either directly (for example, when a register identifier is in the microinstruction) or indirectly (for example, when the microinstruction indicates that an instruction register contains the register identifier). Because of dependencies within the temporary microinstruction state, there are many more dependencies between instructions than just appear at the program level, and these dependencies arise in a greater variety of ways.

The renaming and recovery mechanisms, if implemented, must understand both of these levels of program state and the various ways they can be accessed. For example, the decode phase of an instruction might apply renaming to the instruction-level registers, passing the renamed identifiers to the microcode level. At the microcode level, we might need an additional renaming mechanism to rename the microcode registers. If recovery is implemented using a structure such as the reorder buffer, it is somewhat difficult for this structure to detect instruction boundaries, because there can be a large number of register and memory updates between instructions. Detecting these boundaries is much easier with a RISC instruction set, because each instruction, in general, updates a single item of state.

This approach is not likely to be very fruitful. It introduces a number of complications, and it is not clear how it can help the hardware discover many independent operations. Out-of-order issue probably does not benefit the microinstruction sequence for a single instruction, because this sequence typically contains mostly dependent operations. We must interleave the microcode sequences of different instructions to uncover instruction parallelism. However, to uncover this parallelism, we must design the hardware to deal with a very large number of microinstructions, because the instruction parallelism shows up at widely separated points within the microinstruction stream.

A.2.2 Overlapping Microinstruction Sequences

When we try to overlap complex instructions by issuing the corresponding micro-instructions out of order, we encounter a problem that we have already examined in this book: the hardware is not uncovering parallelism because it is fetching mostly dependent operations–that is, dependent microinstructions. We have seen that the solution to this problems is software scheduling. In the case of a RISC processor, software might reorder instructions so that the hardware fetches independent operations, but we cannot apply this technique directly to the CISC microprocessor. To implement a superscalar CISC processor, we can use a variation of this idea at the microcode level.

Adams and Zimmerman [1989] report that, in the 8086 processor, 25 instructions account for over 90% of the execution of popular personal-computer programs. Although there are no published measurements on the 386, it is likely that it also executes a relatively small portion of its instruction repertoire most of the time. This is the very realization that motivated RISC architectures in the first place. This phenomenon also should hold in the future, even as new applications are developed, because new applications probably are going to be developed in high-level languages. Compilers typically generate instructions in a stylized fashion, using a subset of the instructions available in a CISC architecture, because code generators often cannot recognize cases where complex instructions can be used.

Because the 386 probably executes a small portion of the instruction set most of the time, there is likely to be a frequent set of instruction *combinations* as well. For example, let us say that the case with the 8086 also holds for the 386, so that 25 of the 386 instructions account for 90% of execution. If we examine these instructions in pairs, we are dealing with 625 pairs. This is a small enough sample to study in an attempt to find instruction pairs that can be interpreted concurrently by microcode, with the effect that the instructions are executed at the same time.

For example, consider two instructions that load operands from the run-time stack in memory. Though it would be difficult to execute these instructions at *exactly* the same time, hardware could execute them at *almost* the same time, even if the addressing modes are complex. If the hardware is aware of the second instruction, it can begin executing this instruction immediately after it begins the first, executing the second instruction in a pipelined fashion with the first and finishing the second instruction one cycle after the first (in the best case). Thus, two instructions that might take three cycles each in a scalar processor, for a total of six cycles, might take a total of four cycles in the superscalar processor. Of course, the performance gained is very dependent on how frequent certain combinations are and how effectively they can be interleaved. Recall, however, that the primary design goal is to show some performance advantage over the scalar processor, not to show an advantage over many other architectures.

A similar technique to the one proposed here is used in the Tandem, Inc., Cyclone processor [Horst et al. 1990]. Figure A-2 shows, conceptually, how instructions are paired. The instruction fetcher supplies instructions to a queue, which decodes instructions at least to the extent of recognizing instruction boundaries and locating two instructions. Once the instructions have been located, the hardware forms, for each instruction, an index into a *pairing table*. This index is unique for each instruction in the instruction set, much as an entry point into microcode would be unique for each instruction. The pairing table provides an entry for each instruction, and each entry contains two fields. The first field contains information that, when compared with the information in the same field for the other instruction, provides an indication of whether or not the two instructions can be overlapped, or *paired*. For example, this field might indicate resource utilization and dependencies. The second field provides half of an entry address into a microcode routine. If the instructions can be paired, the two entry-point fields are concatenated, resulting in an entry point to a microcode routine that executes both instructions at the same time. Note that, since the entry point is obtained from two different instructions, there is a sufficient number of entry points to execute every possible pair. Of course, not all pairs of instructions can be overlapped. A nice feature of this approach is that the nonpairable instructions–and the very complex instructions–can be executed by traditional, straightforward microcode interpretation.

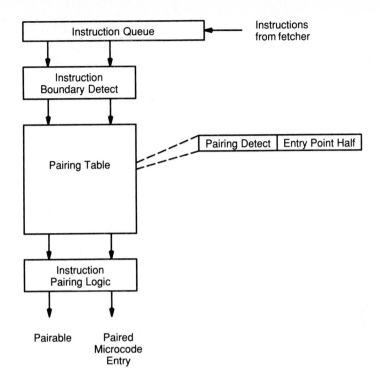

Figure A-2. Pairing Instructions for Overlapped Execution

This approach has a significant difficulty, however. While 625 combinations of instructions is not a very large number of combinations, it is not a very small number either. Pairing instructions introduces a large expansion in microcode–we have introduced as many as 625 new microcode routines. The space taken in the control store for these extra routines is probably not very important. The problem is the other costs associated with these routines: identifying pairable instructions, designing the coding of the pairing table, designing the microinstruction format to allow pairing, verifying that the microcode is correct for each pair, and so on. Furthermore, this technique does not have much of a future. Even if it is manageable to pair instructions, it is certainly hopeless to execute instructions in triples or quadruples.

A.2.3 Superscalar Execution of a "RISC Core" Instruction Set

An approach that has more promise than the ones described in the preceding sections defines the hardware control and data paths to achieve parallelism in specific–but important–cases. We saw in the preceding section that the 8086 processor executes a small number of its instructions most of the time. These instructions are very much like common RISC instructions. Indeed, Intel has relied on this observation to speed up the execution of frequent instructions in the i486, using pipelining to achieve an execution rate they claim is nearly one instruction per cycle for common instructions.

What is needed for the superscalar implementation is a simple extension of this concept: the superscalar execution of common, single-cycle instructions. It should be possible to define a core set of instructions that have at least some "RISC-like" qualities, such as having register identifiers in only a few places, using only a few addressing modes, having only a few different instruction lengths, operating on fixed-width registers, and so on. Once the datapaths have been designed to execute these instructions in a single cycle, we are close to being able to apply the techniques we have studied in this book. Of course, there always must be some provision for the most complex instructions and other instructions not in the core set. These can be handled by serializing the processor, then executing the instructions using traditional microcode techniques.

This approach has the advantage that it paves the way not only for the hardware techniques that we have examined, but also for the software techniques. With the definition of a core set of instructions that the hardware executes efficiently, it is possible to develop compilers and schedulers that focus on this core set. Having a core set of important, efficient instructions reduces software scheduling for a superscalar 386 to a manageable task. Even though it is not possible to recompile or schedule every 386 application, the ability to do so provides a path for future performance enhancement.

A.3 CONCLUSION

We have seen that, though the 386 is complex and a superscalar implementation is difficult, there is enough economic motivation and a sufficiently low risk that a superscalar version probably will be implemented. Moreover, the design is not impossibly complex. The designers can follow the same path followed by the i486: accelerate the common cases that are most frequently executed. The architecture must carry along baggage for compatibility reasons, but this only increases the cost of— and delays the introduction of–the design. Because the 386 has exclusive access to the personal-computer market, these costs and delays are secondary considerations.

References

[AMD 1989] Advanced Micro Devices, Inc. *Am29000 Users Manual*, No. 10620B (1989). Sunnyvale, CA.

[ASC/X3 1988] Accredited Standards Committee X3, Information Processing Systems. *Draft Proposed American National Standard for Information Systems–Programming Language C*, Doc. No. X3J11/88-159. Washington, DC: Computer and Business Equipment Manufacturers Association, December 1988.

[Acosta et al. 1986] Acosta, R. D., J. Kjelstrup, and H. C. Torng. "An Instruction Issuing Approach to Enhancing Performance in Multiple Functional Unit Processors." *IEEE Transactions on Computers*, Vol. C-35 (September 1986), pp. 815-828.

[Adam et al. 1974] Adam, T. L., K. M. Chandy, and J. R. Dickson. "A Comparison of List Schedules for Parallel Processing Systems." *Communications of the ACM*, Vol. 17 (December 1974), pp. 685-690.

[Adams and Zimmerman 1989] Adams, T.L., and R. E. Zimmerman. "An Analysis of 8086 Instruction Set Usage in MS DOS Programs." *Proceedings of the Third International Conference on Architectural Support for Programming Languages and Operating Systems* (April 1989), pp. 152-160.

[Agerwala and Cocke 1987] Agerwala, T., and J. Cocke. "High Performance Reduced Instruction Set Processors," Technical Report RC12434 (#55845). Yorktown, NY: IBM Thomas J. Watson Research Center, January 1987.

[Aho et al. 1986] Aho, A.V., R. Sethi, and J. D. Ullman. *Compilers: Principles, Techniques, and Tools*. Reading, MA: Addison-Wesley, 1986.

[Aiken and Nicolau 1988] Aiken, A., and A. Nicolau. "Perfect Pipelining: A New Loop Parallelization Technique." In H. Ganzinger (ed.) *Proceedings of the 2nd European Symposium on Programming*, pp. 221-235. New York: Springer-Verlag, March 1988.

[Apollo 1988] Apollo Computer, Inc. *The Series 10000 Personal Supercomputer* (1988). Chelmsford, MA.

[Auslander and Hopkins 1982] Auslander, M., and M. Hopkins. "An Overview of the PL.8 Compiler." *Proceedings of the SIGPLAN '82 Symposium on Compiler Construction* (June 1982), pp. 22-31.

[Backus 1978] Backus, J. "Can Programming Be Liberated from the Von Neumann Style? A Functional Style and Its Algebra of Programs." *Communications of the ACM*, Vol. 21, no. 8 (August 1978), pp. 613-641.

[Chaitin et al. 1981] Chaitin, G. J., M. A. Auslander, A. K. Chandra, J. Cocke, M. E. Hopkins, and P. W. Markstein. "Register Allocation via Coloring." *Computer Languages*, Vol. 6, no. 1 (1981), pp. 47-57.

[Charlesworth 1981] Charlesworth, A.E., "An Approach to Scientific Array Processing: The Architectural Design of the AP–120B/FPS 164 Family." *Computer*, Vol. 14 (September 1981), pp. 18-27.

[Chow 1983] Chow, F., *A Portable Machine-Independent Global Optimizer–Design and Measurements*, Ph.D. dissertation. Stanford University, Stanford, CA (December 1983).

[Cohn et al. 1989] Cohn, R., T. Gross, M. Lam, and P. S. Tseng. "Architecture and Compiler Tradeoffs for a Long Instruction Word Microprocessor." *Proceedings of the Third International Conference on Architectural Support for Programming Languages and Operating Systems* (April 1989), pp. 2-14.

[Colwell et al. 1987] Colwell, R. P., R. P. Nix, J. J. O'Donnel, D. B. Papworth, and P. K. Rodman. "A VLIW Architecture for a Trace Scheduling Compiler." *Proceedings of the Second International Conference on Architectural Support for Programming Languages and Operating Systems* (October 1987), pp. 180-192.

[Davidson et al. 1981] Davidson, S., D. Landskov, B. D. Shriver, and P. W. Mallett. "Some Experiments in Local Microcode Compaction for Horizontal Machines." *IEEE Transactions on Computers*, Vol. C-30, no. 7 (July 1981), pp. 460-477.

[Dennis and Misunas 1975] Dennis, J. B. and D. P Misunas. "A Preliminary Architecture for a Basic Data Flow Processor." *Proceedings of the Second International Symposium on Computer Architecture* (January 1975), pp. 126-132.

[Dijkstra 1965] Dijkstra, E. W. "Solution of a Problem in Concurrent Programming Control." *Communications of the ACM*, Vol. 8, no. 9 (September 1965), p. 569.

[Ditzel and McLellan 1987] Ditzel, D.R., and H.R. McLellan. "Branch Folding in the CRISP Microprocessor: Reducing Branch Delay to Zero." *Proceedings of the 14th Annual Symposium on Computer Architecture* (June 1987), pp. 2-9.

[Dubois et al. 1986] Dubois, M., C. Scheurich, and F. A. Briggs. "Memory Access Buffering in Multiprocessors." *Proceedings of the 13th International Symposium on Computer Architecture* (June 1986), pp. 434-443.

[Dwyer and Torng 1987] Dwyer, H., and H. C. Torng, *A Fast Instruction Dispatch Unit for Multiple and Out-of-Sequence Issuances*. School of Electrical Engineering Technical Report EE-CEG-87-15 (November 1987), Cornell University, Ithaca, NY.

[Ellis 1986] Ellis, J. R., *Bulldog: A Compiler for VLIW Architectures*. Cambridge, MA: MIT Press, 1987.

[Fisher 1979] Fisher, J. A. *The Optimization of Horizontal Microcode Within and Beyond Basic Blocks: An Application of Processor Scheduling with Resources*, Ph.D. dissertation, Technical

Report COO-3077-161. Courant Mathematics and Computing Laboratory, New York University, New York (October 1979).

[Fisher 1981] Fisher, J. A. "Trace Scheduling: A Technique for Global Microcode Compaction." *IEEE Transactions on Computers*, Vol. C-30 (July 1981), pp. 478-490.

[Fisher 1983] Fisher, J. A. "Very Long Instruction Word Architectures and the ELI-512." *Proceedings of the 10th Annual Symposium on Computer Architectures* (June 1983), pp. 140-150.

[Foster and Riseman 1972] Foster, C. C. and E. M. Riseman. "Percolation of Code to Enhance Parallel Dispatching and Execution." *IEEE Transactions on Computers*, Vol. C-21 (December 1972), pp. 1411-1415.

[Goodman and Hsu 1988] Goodman, J. R. and W.-C. Hsu. "Code Scheduling and Register Allocation in Large Basic Blocks." *Proceedings of the 1988 International Conference on Supercomputing* (July 1988), pp. 442-452.

[Gross et al. 1988] Gross, T. R., J. L. Hennessy, S. A. Przybylski, and C. Rowen. "Measurement and Evaluation of the MIPS Architecture and Processor." *ACM Transactions on Computer Systems*, Vol. 6, no. 3 (August 1988), pp. 229-257.

[Gross and Hennessy 1982] Gross, T. R., and J. L. Hennessy. "Optimizing Delayed Branches." *Proceedings of IEEE Micro-15* (October 1982), pp. 114-120.

[Groves and Oehler 1989] Groves, R. D., and R. Oehler. "An IBM Second Generation RISC Processor Architecture." *Proceedings of the 1989 IEEE International Conference on Computer Design: VLSI in Computers and Processors* (October 1989), pp. 134-137.

[Hennessy 1986] Hennessy, J. L. "RISC-Based Processors: Concepts and Prospects." *New Frontiers in Computer Architecture Conference Proceedings* (March 1986), pp. 95-103.

[Hennessy and Gross 1983] Hennessy, J., and T. Gross. "Postpass Code Optimization of Pipeline Constraints." *ACM Transactions on Programming Languages and Systems*, Vol. 5, no. 3 (July 1983), pp. 422-448.

[Horst et al. 1990] Horst, R. W., R. L. Harris, and R. L. Jardine. "Multiple Instruction Issue in the NonStop Cyclone Processor." *Proceedings of the 17th Annual Symposium on Computer Architecture* (May 1990), pp. 216-226.

[Hwu and Chang 1988] Hwu, W., and P. P. Chang. "Exploiting Parallel Microprocessor Microarchitectures with a Compiler Code Generator." *Proceedings of the 15th Annual Symposium on Computer Architecture* (June 1988), pp. 45-53.

[Hwu and Patt 1986] Hwu, W., and Y. N. Patt. "HPSm, a High Performance Restricted Data Flow Architecture Having Minimal Functionality." *Proceedings of the 13th Annual Symposium on Computer Architecture* (June 1986), pp. 297-307.

[Hwu and Patt 1987] Hwu, W.-M., and Y. N. Patt. "Checkpoint Repair for Out-of-Order Execution Machines." *Proceedings of the 14th Annual Symposium on Computer Architecture* (June 1987), pp. 18-26.

[Intel 1986] Intel Corporation. *80386 Programmer's Reference Manual*, No. 230985–001 (1986). Santa Clara, CA.

[Intel 1989a] Intel Corporation. *i860 64-Bit Microprocessor Programmer's Reference Manual*, No. 240329–001 (1989). Santa Clara, CA.

[Intel 1989b] Intel Corporation. *80960CA User's Manual*, No. 270710–001 (1989). Santa Clara, CA.

[Johnson 1989] Johnson, W. M. *Super-Scalar Processor Design*, Ph.D. dissertation, Technical Report CSL-TR-89-383. Stanford University, Stanford, CA (June 1989).

[Jouppi 1989a] Jouppi, N. P. "Integration and Packaging Plateaus of Processor Performance." *Proceedings of the 1989 IEEE International Conference on Computer Design: VLSI in Computers and Processors* (October 1989), pp. 229-232. Also published as Research Report 89/10, Digital Equipment Corporation Western Research Laboratory, Palo Alto, CA (July 1989).

[Jouppi 1989b] Jouppi, N. P. "The Distribution of Instruction-Level and Machine Parallelism and Its Effect on Performance." *IEEE Transactions on Computers*, Vol. 38, no. 12 (December 1989), pp. 1645-1658. Also published as Research Report 89/13, Digital Equipment Corporation Western Research Laboratory, Palo Alto, CA (July 1989).

[Jouppi and Wall 1988] Jouppi, N. P., and D. W. Wall. "Available Instruction-Level Parallelism for Superscalar and Superpipelined Machine." *Proceedings of the Third International Conference on Architectural Support for Programming Languages and Operating Systems* (April 1989), pp. 272-282. Also published as Technical Note TN-2, Digital Equipment Corporation Western Research Laboratory, Palo Alto, CA (September 1988).

[Kane 1987] Kane, G., *MIPS R2000 RISC Architecture*. Englewood Cliffs, NJ: Prentice-Hall, 1987.

[Keller 1975] Keller, R. M. "Look-Ahead Processors." *Computing Surveys*, Vol. 7, no. 4 (December 1975), pp. 177-195.

[Kogge 1981] Kogge, P. M. *The Architecture of Pipelined Computers*. New York: McGraw-Hill, 1981.

[Lam 1988] Lam, M. "Software Pipelining: An Effective Scheduling Technique for VLIW Machines." *Proceedings of the SIGPLAN '88 Conference on Programming Language Design and Implementation* (June 1988), pp. 318-328.

[Lam 1989] Lam, M. *A Systolic Array Optimizing Compiler*. Boston: Kluwer, 1989.

[Lam 1990] Lam, M. S. "Instruction Scheduling for Superscalar Architectures." *Annual Review of Computer Science*, Vol. 4 (1990), pp. 173-201.

[Lee and Smith 1984] Lee, J. K. F., and A. J. Smith. "Branch Prediction Strategies and Branch Target Buffer Design." *IEEE Computer*, Vol. 17 (January 1984), pp. 6-22.

[MIPS 1986] MIPS Computer Systems, Inc. *MIPS Language Programmer's Guide* (1986). Sunnyvale, CA.

[MIPS 1989] MIPS Computer Systems, Inc. *Performance Brief* (June 1989). Sunnyvale, CA.

[McFarling and Hennessy 1986] McFarling, S., and J. Hennessy. "Reducing the Cost of Branches." *Proceedings of the 13th Annual Symposium on Computer Architecture* (June 1986), pp. 396-404.

[Motorola 1989] Motorola, Inc. *MC88100 RISC Processor User's Manual*. Englewood Cliffs, NJ: Prentice-Hall, 1989.

[Murakami et al. 1989] Murakami, K., N. Irie, M. Kuga, and S. Tomita. "SIMP (Simple Instruction stream/Multiple instruction Pipelining): A Novel High-Speed Single-Processor Architecture." *Proceedings of the 16th Annual Symposium on Computer Architecture* (June 1989), pp. 78-85.

[Nicolau 1985] Nicolau, A. *Percolation Scheduling: A Parallel Compilation Technique*, Department of Computer Science Technical Report TR 85-678. Cornell University, Ithaca, NY (May 1985).

[Nicolau and Fisher 1984]   Nicolau, A., and J. A. Fisher. "Measuring the Parallelism Available for Very Long Instruction Word Architectures." *IEEE Transactions on Computers*, Vol. C-33 (November 1984), pp. 968-976.

[Patt et al. 1985a]   Patt, Y. N., W. Hwu, and M. Shebanow. "HPS, A New Microarchitecture: Rationale and Introduction." *Proceedings of the 18th Annual Workshop on Microprogramming* (December 1985), pp. 103-108.

[Patt et al. 1985b]   Patt, Y. N., S. W. Melvin, W. Hwu, and M. C. Shebanow. "Critical Issues Regarding HPS, A High Performance Microarchitecture." *Proceedings of the 18th Annual Workshop on Microprogramming* (December 1985), pp. 109-116.

[Pleszkun et al. 1987]   Pleszkun, A., J. Goodman, W. C. Hsu, R. Joersz, G. Bier, P. Woest, and P. Schecter. "WISQ: A Restartable Architecture Using Queues." *Proceedings of the 14th Annual Symposium on Computer Architecture* (June 1987), pp. 290-299.

[Pleszkun and Sohi 1988]   Pleszkun, A. R., and G. S. Sohi. "The Performance Potential of Multiple Functional Unit Processors." *Proceedings of the 15th Annual Symposium on Computer Architecture* (June 1988), pp. 37-44.

[Przybylski et al. 1988]   Przybylski, S., M. Horowitz, and J. Hennessy. "Performance Tradeoffs in Cache Design." *Proceedings of the 15th Annual Symposium on Computer Architecture* (June 1988), pp. 290-298.

[Rau and Glaeser 1981]   Rau, B. R., and C. D. Glaeser. "Some Scheduling Techniques and an Easily Schedulable Horizontal Architecture for High Performance Scientific Computing." *Proceedings of the 14th Annual Workshop on Microprogramming* (October 1981), pp. 183-198.

[Riseman and Foster 1972]   Riseman, E. M., and C. C. Foster. "The Inhibition of Potential Parallelism by Conditional Jumps." *IEEE Transactions on Computers*, Vol. C-21 (December 1972), pp. 1405-1411.

[Sohi and Vajapeyam 1987]   Sohi, G. S., and S. Vajapeyam. "Instruction Issue Logic for High-Performance Interruptable Pipelined Processors." *Proceedings of the 14th Annual Symposium on Computer Architecture* (June 1987), pp. 27-34.

[Smith 1982]   Smith, A. J. "Cache Memories." *ACM Computing Surveys*, Vol. 14, no. 3 (September 1982), pp. 473-530.

[Smith et al. 1989]   Smith, M. D., M. Johnson, and M. A. Horowitz. "Limits on Multiple Instruction Issue." *Proceedings of the Third International Conference on Architectural Support for Programming Languages and Operating Systems* (April 1989), pp. 290-302.

[Smith et al. 1990]   Smith, M. D., M. S. Lam, and M. A. Horowitz. "Boosting Beyond Static Scheduling in a Superscalar Processor." *Proceedings of the 17th Annual Symposium on Computer Architecture* (May 1990), pp. 344-354.

[Smith and Pleszkun 1985]   Smith, J. E., and A. R. Pleszkun. "Implementation of Precise Interrupts in Pipelined Processors." *Proceedings of the 12th Annual International Symposium on Computer Architecture* (June 1985), pp. 36-44.

[Thorton 1970]   Thorton, J. E. *Design of a Computer–The Control Data 6600*. Glenview IL: Scott, Foresman and Co., 1970.

[Tjaden 1972]   Tjaden, G. S. *Representation and Detection of Concurrency using Ordering Matrices*, Ph.D. dissertation. The Johns Hopkins University, Baltimore, MD (1972).

[**Tjaden and Flynn 1970**]   Tjaden, G. S., and M. J. Flynn. "Detection and Parallel Execution of Independent Instructions." *IEEE Transactions on Computers*, Vol. C-19, no. 10 (October 1970), pp. 889-895.

[**Tjaden and Flynn 1973**]   Tjaden, G. S., and M. J. Flynn. "Representation of Concurrency with Ordering Matrices." *IEEE Transactions on Computers*, Vol. C-22, no. 8 (August 1973), pp. 752-761.

[**Tomasulo 1967**]   Tomasulo, R.M. "An Efficient Algorithm for Exploiting Multiple Arithmetic Units." *IBM Journal*, Vol. 11 (January 1967), pp. 25-33.

[**Torng 1984**]   Torng, H. C., *An Instruction Issuing Mechanism for Performance Enhancement.* School of Electrical Engineering Technical Report EE-CEG-84-1, Cornell University, Ithaca, NY (February 1984).

[**Uht 1986**]   Uht, A. K. "An Efficient Hardware Algorithm to Extract Concurrency from General-Purpose Code." *Proceedings of the Nineteenth Annual Hawaii International Conference on System Sciences* (1986), pp. 41-50.

[**Uvieghara et al. 1990**]   Uvieghara, G. A., Y. Nakagome, D.-K. Jeong, and D. A. Hodges. "An On-Chip Smart Memory for a Data-Flow CPU." *IEEE Journal of Solid-State Circuits*, Vol. 25, no. 1 (February 1990), pp. 84-94.

[**Waterside 1989**]   Waterside Associates. "Benchmark Results." *SPEC Newsletter*, Vol. 1, no. 1 (Fall 1989).

[**Wedig 1982**]   Wedig, R. G. *Detection of Concurrency in Directly Executed Language Instruction Streams*, Ph.D. dissertation. Stanford University, Stanford, CA (June 1982).

[**Weiss and Smith 1984**]   Weiss, S., and J. E. Smith. "Instruction Issue Logic in Pipelined Supercomputers." *IEEE Transactions on Computers*, Vol. C-33 (November 1984), pp. 1013-1022.

[**Weiss and Smith 1987**]   Weiss, S., and J. E. Smith. "A Study of Scalar Compilation Techniques for Pipelined Supercomputers." *Proceedings of the Second International Conference on Architectural Support for Programming Languages and Operating Systems* (October 1987), pp. 105-109.

[**Wulf 1988**]   Wulf, W. A. "The WM Computer Architecture." *Architecture News* (January 1988), pp. 70-84.

Index

A

Acosta et al. [1986], 55, 113, 135

Adam et al. [1974], 193

Adams and Zimmerman [1989], 265, 269

Agerwala and Cocke [1987], 29, 55

Aho et al. [1986], 182

Aiken and Nicolau [1988], 221

AMD [1989], 19, 178

Am29000 processor, 178, 189, 204

ANSI C, 190, 246

Apollo [1988], 250

ASC/X3 [1988], 190

Auslander and Hopkins [1982], 5, 180

addressing. *See* loads/stores, addressing

aliasing
   of program variables, 151
   of virtual addresses, 52

aligning instruction runs
   hardware aligning, 66
   performance benefit, 67-69

software aligning, 69

antidependencies. *See* dependencies, anti-

application-specific processors, 245-246

arbitration
   operands buses, 144-145
   register ports, 79-81, 168
     serial arbiter, 81
   resource, 145
   result buses, 35, 120-122

architectural state, definition, 89

arithmetic mean. *See* harmonic mean,
   comparison to arithmetic and geometric
   mean

associative lookup, definition, 49

B

Backus [1978], 23

backtracking, in software scheduling, 199

basic block, definition, 182

benchmark programs, 35
   sample benchmarks, 37, 46, 59, 63, 65, 78,
     97, 120, 122

boosting, 243

branch block index field, 73, 74-75

branch delay
definition, 58
in scalar processor, 62
in superscalar processor, 58-59, 62-63

branch prediction, 57, 63-65, 171-172
See also recovery
hardware implementation, 63, 71-76
hardware versus software, 64, 85
indirect branches, 72
penalty of misprediction, 97-98
performance benefit, 67-69
recovery using future file, 96-97
recovery using reorder buffer, 96
software implementation, 63

branch-and-bound scheduler, 198

branch-target buffer, 63, 71-72

branches, 12-13, 54-55, 57-86, 131
alignment penalty of, 58, 59, 65-66
See also aligning instruction runs;
merging instruction runs
data-dependent branches and software
scheduling, 214-216, 229
delay penalty of, 57-58
See also branch prediction
removing by multiple-path execution,
69-71
delayed, 61-63, 204
hardware implementation, 81-84
number pending, 81-82
order of execution, 75, 82-83
target address computation, 83-84

buses, performance effects, 79, 120, 134,
143-144

C

Chaitin et al. [1981], 23

Charlesworth [1981], 218, 221

Chow [1983], 5

CISC processor, 163, 191
See also Intel 386 architecture; RISC
processor, compared to CISC
definition, 3-4
procedural dependencies in, 13
resource conflicts in, 14
true dependencies in, 11

CMOS, 248

Cohn et al. [1989], 26, 240

Colwell et al. [1987], 26, 205

caches, 4, 33, 76, 247, 248-250, 251-252,
255-256
data, 43-44, 52-53, 159-161, 163
instruction, branch prediction by, 72-76
noncacheable accesses, 153-154
trade-offs with superscalar hardware, 254

central window, 128-129, 133-145
See also reservation stations, compared to
central window
operand buses, 46-47, 142-144

checkpoint repair, 88, 89-91
disadvantages, 91

clock skew, 27-28, 249

compaction, 25-26

compensation code. *See* software scheduling,
trace scheduling

compiler optimization, 175
See also software scheduling
compaction, 25-26
effect on instruction parallelism, 32
register allocation, 23, 109-110
software scheduling interaction, 187-189

complex-instruction-set processor. *See* CISC
processor

content-addressable memory, definition, 49

control flow graph, 182

critical path
in code, 186-187, 192
in hardware, 69

cycle time, 175, 258
definition, 2

cycle-time margin, 253-254, 255-256,
    257-258
  definition, 253

D

Davidson et al. [1981], 192, 197

Dennis and Misunas [1985], 104

Dijkstra [1965], 162

Ditzel and McLellan [1987], 64, 71

Dubois et al. [1986], 162

Dwyer and Torng [1987], 55, 135, 137

data cache. *See* caches, data

data parallelism, 29-30, 103-104

dataflow architectures, 29, 104

dataflow graph, 184, 192

deadlock
  in load/store unit, 154
  in software scheduling, 199

decode-to-issue delay, 139

decoder, 168-170
  branch instructions, 83-84
  four-instruction, 77-81, 171-172
    See also decoder, two-instruction versus
      four-instruction
  two-instruction versus four-instruction, 59,
    65-66, 69, 85

degree of superpipelining, 26

dependencies, 20, 22, 35, 168
  anti-, 111, 112-113, 186, 188
    avoiding by copying, 115-116
    definition, 22
  between loads and stores, 50-53, 151, 154,
    157-158, 162-163
  interlocks, 112-115
    definition, 107
  loop-carried, 223-224
  output, 21, 111, 113, 186, 188
    definition, 20
    enforcing via antidependencies, 115

procedural, 9, 12-13, 38, 39, 54-55
    See also branches
    CISC processor and, 13
    definition, 12
    software versus hardware enforcing,
      105-107, 228
  true, 9, 10-12, 14, 19, 21, 26-27, 39-41
    CISC processor and, 11
    definition, 10
    implied operands, 11

design-time margin, 253-254, 255-256,
    257-258
  definition, 253

direct tag search, 123-124

dispatch stack, 113-115, 129, 135-137
  dependency mechanism, compared to
    renaming, 114
  effect on performance, 135
  hardware costs, 113-114, 135-137

E

ECL, 248

Ellis [1986], 205

epilogue, in software pipelining. *See* software
  scheduling, software pipelining

exception restart. *See* restart

exceptions, 20

F

Fisher [1979], 193, 205

Fisher [1981], 25, 205

Fisher [1983], 25, 205

Foster and Riseman [1972], 230

fetch efficiency, 65

first-come, first-served scheduler, 197

floating-point, 53, 147, 229, 257
  effect on instruction parallelism, 39-40

flow dependencies. *See* dependencies, true

forwarding, 122-125
  pipeline hazard, 169-170
  store data, 151

four-instruction decoder. *See* decoder,
    four-instruction; decoder, two-instruction
    versus four-instruction

frequency scaling, 247-248

future file, 88, 125-126
  hardware costs, 111-112
  operation of, 94-95
  renaming implementation, 110-112

G

Goodman and Hsu [1988], 189

Gross and Hennessy [1982], 61

Gross et al. [1988], 43, 51, 60

Groves and Oehler [1989], 55, 262

geometric mean. *See* harmonic mean,
    comparison to arithmetic and geometric
    mean

greedy scheduler, 197-198

H

Hennessy [1986], 5

Hennessy and Gross [1983], 35, 105, 180,
    181, 189

Horst et al. [1990], 270

HPS (high performance substrate), 268

Hwu and Chang [1988], 23

Hwu and Patt [1986], 15, 55

Hwu and Patt [1987], 89

harmonic mean, 40
  compared to arithmetic and geometric
      mean, 37
  definition, 36-37

heuristics, 191

hierarchical reduction. *See* software
    scheduling, global code motion

history buffer, 88, 91-92, 210
  disadvantages, 92

horizontal microinstructions, 205

hypercubes, 29

I

IBM 3033, 70

IBM 370/Model 168, 70

IBM RIOS architecture, 55, 262

IBM RISC System/6000, 55

Intel [1986], 261

Intel [1989a], 240, 262

Intel [1989b], 108, 240

Intel 386 architecture
  complex instructions in, 266-267
  instruction format, 262-264
  memory dependencies in, 265-266
  register dependencies in, 264-265
  superscalar implementation, 261-272

Intel 80386 processor, 261, 267, 268

Intel 8086 processor, 265, 269-270, 271

Intel 80960CA processor, 108, 240-241

Intel i486 processor, 261, 267-268, 272

Intel i860 processor, 240, 262

in-order completion, 18, 53, 88, 89, 102

in-order issue, 18, 19-20, 34, 53

in-order state, 50
  definition, 88-89

initiation interval, in software pipelining. *See*
    software scheduling, software pipelining

instruction cache. *See* caches, instruction

instruction issue, 17
  See also in-order issue; instruction-issue
      policy; out-of-order issue

instruction parallelism, 17, 103-104
  definition, 15

limits, 38-41

instruction runs
    definition, 58
    software scheduling benefit and, 181

instruction set design, for superscalar
    processor, 239-244

instruction trace, 54
    in simulation, 32, 33-34, 38
    in trace scheduling, 205

instruction window, 46, 173-174, 199-200
    See also central window; reservation
        stations
    definition, 21
    operation, 21

instruction-issue policy, 17-24, 34-35
    See also in-order issue; out-of-order issue
    definition, 17

interlocks, 123
    dependency interlocks. *See* dependencies,
        interlocks
    process interlocks, 162

issue latency, definition, 32

J

Johnson [1989], 55

Jouppi [1989a], 55, 248

Jouppi [1989b], 11, 16, 26

Jouppi and Wall [1988], 26, 55

K

Kane [1987], 32

Keller [1975], 17, 23, 50

Kogge [1981], 3, 28

L

Lam [1988], 218, 221, 231

Lam [1989], 219, 221, 231

Lam [1990], 182

Lee and Smith [1984], 63, 70, 71, 76

loads/stores, 131-132, 147-164
    See also store buffer
    addressing, 50, 154-156, 162-163
    bypassing stores, 52, 150-151, 174-175
    deadlock, 154
    dependencies. *See* dependencies, between
        loads and stores
    forwarding, 151
    order of issue, 148-154
    parallelism, 43-44, 159-161
    performance, 152-153, 155-156
    reorder buffer role, 100
    side effects, 153-154
    standard processor mechanism, 50-53
    virtual addressing, 52

logical spaces, in checkpoint repair, 89-90

lookahead, 21, 130
    branch delay effect, 63
    branch prediction and, 64-65
    definition, 17
    in software scheduler, 198-199

lookahead state, definition, 89

loop unrolling. *See* software scheduling, loop
    unrolling

M

McFarling and Hennessy [1986], 12, 62, 64,
    71, 213

MIPS [1986], 33

MIPS [1989], 32

Motorola [1989], 19, 108

Motorola 88100 processor, 108

Murakami et al. [1989], 55

machine parallelism, 17-24
    definition, 16
    limits, 41-44
    performance with infinite parallelism, 40-41

---

memory addressing. *See* loads/stores, addressing

memory technology, 4, 248-249, 255, 256

merging instruction runs
    hardware merging, 66
    performance benefit, 67-69
    software merging, 69

microcode, 268-272
    pairing instruction execution, 270

misalignment. *See* branches, alignment penalty of

modulo reservation table, in software pipelining. *See* software scheduling, software pipelining

multiprocessors, 28, 161-162, 259

N

Nicolau [1985], 230

Nicolau and Fisher [1984], 12

O

operand buses. *See* central window, operand buses; reservation stations, operand buses

operation latency, 186
    definition, 11

optimization. *See* compiler optimization

out-of-order completion, 19-20, 34, 53, 88-89, 102

out-of-order issue, 21-22, 38, 41, 55, 57, 102, 146, 163, 171-172, 175
    compared to software scheduling, 200-202, 233-236
    of microinstructions, 268-269
    standard processor mechanism, 46-48

output dependencies. *See* dependencies, output

P

Patt et al. [1985a], 15, 55, 268

Patt et al. [1985b], 15

Pleszkun and Sohi [1988], 55

Pleszkun et al. [1987], 22, 55, 63, 93

Przybylski et al. [1988], 76

packaging technology, 1, 4, 238, 247, 248-249, 254, 256

pairing table, 270

parallelism. *See* data parallelism; instruction parallelism; machine parallelism

partial renaming. *See* register renaming, partial renaming

path latency, 186

percolation. *See* software scheduling, global code motion

performance
    hardware simplifications and, 172-175
    principles in improving, 2-3

performance growth, 251, 255, 258

pipeline stages
    comparison of scalar and superscalar processors, 50
    definition, 3

pipelining, 50
    See also superpipelining
    definition, 3

pixie, 33

preaddress buffer, 154-156

precedence constraints, 184, 192

precedence graph, 184-186

precedence rules, 190

precise exceptions. *See* precise interrupts

precise interrupts, 22, 49, 111, 137
    definition, 20

predecessors, in flow graph, 183

procedural dependencies. *See* dependencies, procedural

resource conflicts, 9, 13-15, 18, 19, 21, 22, 27, 35, 39, 70
  See also storage conflicts
  CISC processor and, 14
  definition, 13
  removing, 14, 41
  removing by duplicating ALU, 43
  removing by duplicating load/store unit, 43-44, 159-161

resource reservation table, 187

resource utilization, 14, 177
  effect of software scheduling, 199-200

restart, 20, 22, 102
  See also recovery
  definition, 87

result latency, definition, 33

risks, 237
  of hardware techniques, 250-260
    cost-oriented scenario, 254-256
    performance-oriented scenario, 251-254
  of software techniques, 244-247

run length
  average versus median, 60
  definition, 58
  distribution in sample benchmarks, 59-60

S

SIMD processors, 28

Smith [1982], 4, 33, 71

Smith and Pleszkun [1985], 20, 48, 55, 91, 92, 94, 95, 99, 100

Smith et al. [1989], 55

Smith et al. [1990], 55, 242, 244

Sohi and Vajapeyam [1987], 55, 93, 96, 129, 134, 137

scalar processor, 127
  branch delay in, 62
  definition, 6
  outperforming superscalar processor, 53-54, 102, 124-125, 139

procedural dependencies in, 12-13
resource conflicts in, 13-14
simulation model, 34-35
software scheduling, 180-181
true dependencies in, 10-11

scientific computation, 1, 16, 28-30, 210-211, 238

scoreboarding, 125-126
  80960CA processor and, 108
  88100 processor and, 108
  compared to renaming, 109-110
  Thorton's algorithm, 107-108

semaphores, 162

semiconductor technology, 1, 28, 143, 247, 248-249, 256, 258

sequence points, 190

serialization, 119-120

shadow register file, 243-244

single-issue cycles, 39, 46
  definition, 11

software incompatibility, 26, 85, 93, 105-106, 235-236, 238, 244-248

software pipelining. *See* software scheduling, software pipelining

software scheduling, 35, 177-202, 203-236, 242-244
  See also compiler optimization
  backtracking, 199
  benefit of, 178-182
  boosting, 243
  branch boundaries and, 203-236
  branch-and-bound scheduler, 198
  compared to out-of-order issue, 200-202, 233-236
  deadlock in, 199
  definition, 177
  delayed branches and, 61-63
  first-come, first-served, 197
  global code motion, 229-233
    hierarchical reduction, 231-233
    percolation, 230-231
  greedy scheduler, 197-198
  instruction format support, 242-244

U

Uht [1986], 55

Uvieghara et al. [1990], 91

unsafe code motion, 212-213, 227

V

VLIW processor, 24, 25-26, 28, 29, 205, 211, 228, 235, 239-241

vector processors, 1, 28, 29-30

very-long-instruction-word processor. *See* VLIW processor

virtual addressing. *See* loads/stores, virtual addressing

W

Waterside [1989], 32

Wedig [1982], 55

Weiss and Smith [1984], 55, 107, 111, 123, 169

Weiss and Smith [1987], 214

Wulf [1988], 55

weak ordering, 162

write-read dependencies. *See* dependencies, true

write-write dependencies. *See* dependencies, output

Z

zero-issue cycles, 39, 46
definition, 11